# The
# ILLUSTRATED
# ENCYCLOPEDIA OF
# ARTILLERY

**A battery of 5.25in coast defence and anti-aircraft guns at Gibraltar**

# The ILLUSTRATED ENCYCLOPEDIA OF ARTILLERY

*An A-Z guide to artillery techniques and equipment throughout the world*

## Ian V. Hogg

CHARTWELL
BOOKS, INC.

A QUARTO BOOK

Published 1988 by
CHARTWELL BOOKS, INC.,
A Division of BOOK SALES INC.
110 Enterprise Avenue
Secaucus, New Jersey 07094

ISBN 1-55521-310-3

This book was designed and produced by
Quarto Publishing plc
The Old Brewery
6 Blundell Street
London N7 9BH

Senior Editor: Helen Owen
Art Editor: Hazel Edington

Editors: Tessa Rose, John Walter

Designer: Richard Slater

Art Director: Moira Clinch
Editorial Director: Carolyn King

Typeset by Text Filmsetters Ltd and Comproom
Ltd
Manufactured in Hong Kong by Regent
Publishing Services Ltd
Printed by Leefung-Asco Printers Ltd, Hong
Kong

# CONTENTS

# A SHORT HISTORY OF ARTILLERY

DATES FROM 1326 TO THE PRESENT

**Above: A 15th century picture of the Siege of La Rochelle by Louis VIII, showing a very early cannon on two trestles.**

The first positive record of artillery is to be found in the manuscript *De Officiis Regnum* by Walter de Milimete, written in 1325. Although there is no reference in the text, there is an illustration which shows a primitive cannon being fired. The cannon consists of a barrel that is almost in the shape of a vase, with a thick base and bell-mouth, and loaded with an arrow. Later evidence revealed that the arrow was bound with leather to make it fit the bore of the gun, and that the propelling charge must have been gunpowder. A hole was drilled through the top of the gun down into the 'chamber' that held the gunpowder; this hole, or 'vent' was filled with fine powder and the gun was fired by the application of a hot iron to this powder.

The first written statement is found in the archives at Florence, where, in 1326, money was appropriated for the provision of brass cannon and iron balls 'for the defence of the commune, camps and territory of Florence.'

Gunpowder had been known for almost a century by this time, the first mention being in the form of an anagram in a document written by Roger Bacon (1214-94) in 1242. Black powder, as it was then known, was something of a novelty, and it was not to be spoken of openly due to restrictions placed on such subjects by the Church. But some time between 1242 and 1325 somebody, somewhere, had the idea of harnessing the explosive force of black powder to propel something from a tube, and artillery was born. There is no evidence to suggest that the Chinese or Arabs invented the cannon, or that they knew of gunpowder before the 13th century. And the famous engraving showing 'Black Berthold', the mysterious monk of Freiburg, who is said to have invented the gun by accidentally firing a charge of powder in his laboratory, can be

Above: Roger Bacon (c.1214-1294) may not have invented gunpowder, but he was certainly the first man to describe it in writing.

dismissed completely, since this dates the invention at 1380; in any case, there seems to be no proof that Black Berthold existed outside the engraver's imagination.

The first cannon appears to have been cast from bronze and was relatively small, perhaps two or three inches in calibre. The art of bell-casting was well enough advanced for it to be utilized in making these early cannon but, as is usually the case, no sooner was the weapon invented than some people began demanding bigger and better. Other methods were adopted when casting could not produce what was wanted.

Below: 'The Town of Durras under Siege' from Froissart's *Chronicles* (1504). Note the early cannon in the foreground, supported by arms attached to a solid bed.

The easiest method was to take the already-known craft of the cooper and adapt it. Guns were therefore made up by placing strips of iron alongside each other in a circle and then binding them by hoops, in exactly the same way that wooden casks and barrels were made, which is probably why we refer to the operative part of a gun as the barrel. The barrel was then heated and the strips hammered to weld them together, more hoops were added, and then the whole barrel was bound in rope and finally wrapped in leather. Once this system was mastered, size became immaterial, and by 1377 the Duke of Burgundy was using cannon capable of firing a 450lb stone shot, roughly equivalent to a calibre of 22 inches.

The cannon of the day was far from accu-

**Left: Another 16th century siege, with mortars and guns engaged on both sides.**

**Below: Depiction of a siege, c.1530, showing cannon mounted upon wheeled carriages, as well as a simple barrel laid on a wooden bed. Barrels of powder and round shot are also in evidence.**

rate; anyone killed by a shot from such a weapon could count himself extremely unlucky, and the general opinion was that they were good only for terrorizing the opposition by their noise and smoke. But against bigger targets such as castles they were more likely to obtain a hit, and one is entitled to wonder why, with such weapons as 22in cannon around, the castles of the 14th century were not knocked to pieces. The answer lies in the gunpowder.

Gunpowder was then, as now, composed of a mixture of sulphur, saltpetre and charcoal. The proportions in the 14th century were 41 percent saltpetre, 30 percent sulphur and 29 percent charcoal, and it is not to be expected that these were to any great degree of purity. Today the proportions are 75:10:15 and the materials are pure and untainted by foreign matter. The early powder was, therefore, weak. Moreover, it was made by grinding up the three constituents separately into fine powders, then mixing them in the dry state. When loaded into the gun, this mixture became a hard and compact mass that was difficult to ignite and when it did eventually ignite burned in fits and starts. Much of the powder was probably ejected from the barrel unburned or still burning, having done little to produce the vital gas that blew the shot from the gun.

The strain upon the gun was not too great because the powder was weak, but the stone

**A Flemish bronze cannon dated 1607, at the Tower of London.**

ball was not thrown with any considerable force, so it took a lot of battering to make any impression on a well-built castle. An iron ball would have been a better proposition, but since iron is much heavier than stone, it would have demanded a bigger charge of powder to throw an iron ball the same distance, and such a charge would probably have wrecked the gun.

Having got the gun, it was necessary to take it to the war, and for that purpose it was simply lashed to a strong wooden bed with rope and moved around by heaving it on and off a strong cart. Such a weapon was obviously not well suited to sudden tactical manoeuvres, and so the employment of artillery in battle became somewhat formal. Dragged into a suitable place in the forefront of the battle, it would fire one or two shots after which the cavalry would charge and the foot soldiers would follow, and the ensuing mêlée in front of the gun would prevent any more shooting. If the enemy pushed the troops back, there would be no time for the laborious task of loading the cannon back on to its cart and it would be left for the enemy to capture.

Towards the end of the 14th century we have the first indications of wheeled artillery

**A heavy mortar firing a powder-filled explosive bomb against a fortified town; from a Portuguese document of the 16th century.**

capable of accompanying marching troops and moving more rapidly on the battlefield. In 1382 the army of Ghent marched against that of Bruges, and took with them a number of *ribauldequins*, light carts upon which a number of small-calibre guns was lashed. This could be moved into position more easily than the wooden-bed gun on a cart, and the several barrels, firing small stone shot, were more likely to do damage. But they were still unlikely to get off more than one or two shots before the tide of battle overtook them.

The slow business of loading (placing the powder in the bore, loading the ball, priming the vent with powder, then touching off the gun; then washing out the bore with a wet cloth on a pole to remove the powder fouling and make sure that no smouldering residue remained before beginning to reload) came under review very early in the history of the gun, and in the mid-14th century 'cannon of two chambers' were recorded. These were simply a barrel open at both ends and fixed into a wooden bed. The chamber and breech end formed a separate unit that could be loaded with powder and ball and then dropped into place at the rear end of the barrel and locked in place by wedges. The second chamber could be loaded in readiness, and once the first had been fired the wedges were knocked away and the chamber replaced by the loaded one. The first chamber could be sponged and reloaded while the second chamber was being secured and fired, and so a moderate rate of fire could be kept up. This gun became known as the *Peterara* and it was to become fairly common in the next century. It eventually fell from grace because the metal-working methods of the day were not good enough to ensure a sound seal between the chamber and the barrel with the result that gas leaked and flames shot out when the gun was fired, and the leakage got worse with each shot. Moreover, the system was not strong enough or reliable enough for large-calibre guns, and is rarely seen on anything of greater calibre than three inches. For the first two hundred or so years of its existence, artillery was a relatively minor factor in warfare. It was expensive and it had to be procured and maintained in peacetime, as it was too late to start thinking about manufacturing cannon and finding the few trained gunners when war had broken out. Few countries in the Middle Ages had sufficient know-

Petraras in use; the one on the left has its breech piece in place, the other two have the breeches open for loading.

ledge of finance to be able to support such a standing burden; the English Army in France at the start of the Hundred Years War had but 15 guns and 84lb of gunpowder, and 15 guns would not spread very far through France to support the army.

Even so, some technical improvement took place during this otherwise slow period. The first improvement was to the gunpowder. The original gunpowder, called 'serpentine', was a fine powder with a tendency to separate out into its component parts when shaken and vibrated in a cart. This meant that the gunner had to re-mix it before he could use it; a dangerous practice because gunpowder is sensitive to friction. It became common practice to carry the three parts separately and mix them as required, dangerous though this was. Another defect was that the fineness of the powder meant that it was easily affected by damp, and the saltpetre attracted moisture. Keeping powder dry was a constant worry to the gunner of the day.

**Left: An 18th century 18pr cannon at Gibraltar.**

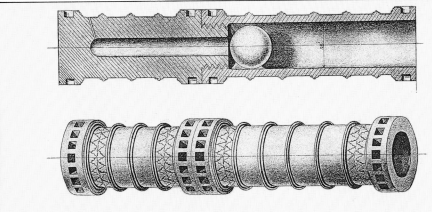

This Turkish cannon is a typical construction of the 15th century. Built to the orders of Mohammed II, its two parts were screwed together.

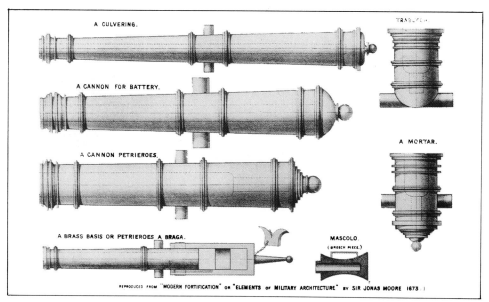

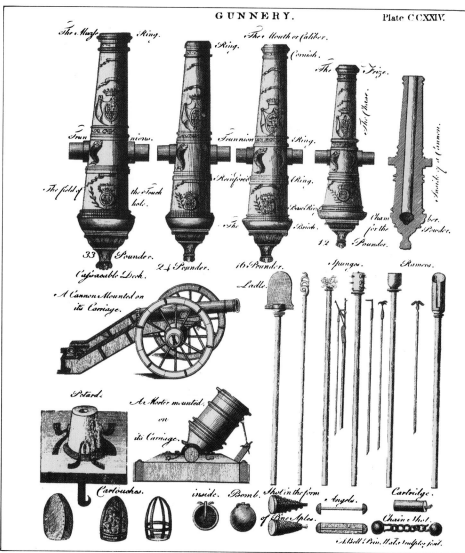

The improvement in gunpowder occurred some time in the early 15th century, and consisted in mixing the three components in a wet state, drying the resulting paste and finally breaking up the 'cake' and passing it through a sieve to obtain grains of a regular size. Called 'corned powder', it was a vast improvement over serpentine. The irregular grains did not pack as tightly when loaded, so that the ignition flame passed through the mass more quickly and fired the powder more certainly, resulting in more efficient ignition and more power. The mixing process made for better incorporation and the grains were less susceptible to damp; the mixture no longer separated during transport, was easier to load, and left less fouling.

The other side of the coin was that the more powerful powder began to burst the old barrel-stave guns, and the complex method of manufacturing the new powder made it extremely expensive. These two factors prevented its acceptance for some time, but they speeded up the development of cannon that would not burst asunder when fired with corned powder. The technique of casting had improved, and it became obvious that this was the only way to produce a strong and practical gun. Bronze guns were cast in France from about 1460 onwards, and cast iron guns were being manufactured in England by 1543. The strength of cast iron guns brought the cast iron cannon-ball into common use.

By the end of the 16th century cast cannon were becoming familiar in all armies, and by this time there was some degree of standardization in calibres and sizes to the point where certain guns were given specific names to identify them. Where these names originated is a mystery, though most of them appear to have been mythical birds. Thus the Robinet was of 1.5in calibre and fired a one-pound shot; the Falcon was 2.5in, firing a 2.5lb shot; the Saker was 3.5in firing a 5lb shot; the Culverin a 5.5in firing an 18lb shot. At the top of the tree was the Cannon, of 8in calibre and firing a 60lb shot.

By the beginning of the 17th century there were numerous books available on the art and science of gunnery, and 'natural philosophers' were beginning to make a serious study of the flight of the shot or, at least, as

**Top: An engraving from 1673 shows the different types of cannon.**

**Left: A page from a 17th century text shows the types of cannon then in use, together with some of their accessories.**

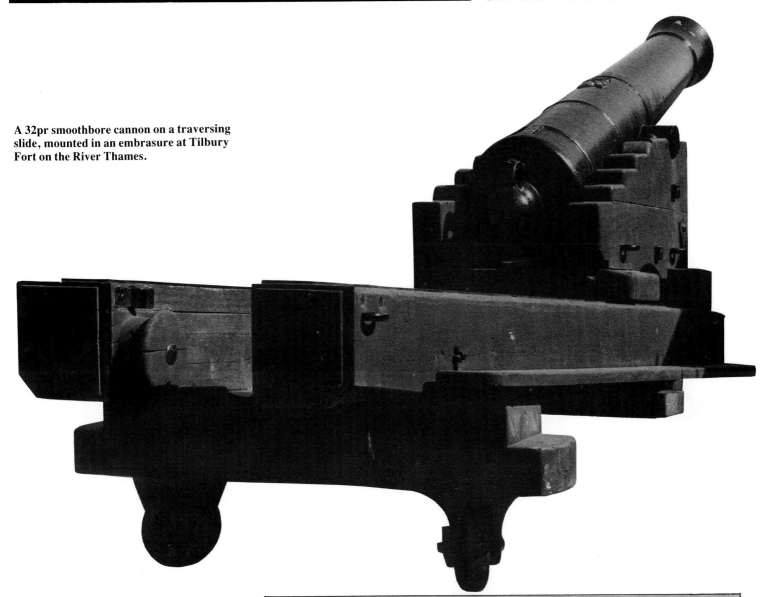

A 32pr smoothbore cannon on a traversing slide, mounted in an embrasure at Tilbury Fort on the River Thames.

serious a study as their knowledge allowed. Some beliefs propagated in print were more akin to superstition; for example, one text recommended the gunner to eat and drink before firing his gun, otherwise the fumes of the powder would be 'hurtful to his brain'; yet in the same text is evidence that by this time the gunner was making up his cartridges in cloth bags of even weight prior to the battle rather than shovelling in scoops of gunpowder estimated by eye. Although in many cases the writers were not prepared to advance theories as to why things happened as they did, from long observation they were able to say that if one gunwheel were higher than the other the shot would deviate from its planned path, and that if the 'air were thick or thinne' the shot would fly a greater or lesser distance than normal.

Right: A 16th century Galloper swivel gun on a four-wheeled carriage, an early form of light artillery which could accompany cavalry.

Mounted upon light two-wheeled carriages these guns were more mobile than anything that had hitherto been seen in battle, even though they had less power and accuracy than a heavier cast iron gun. But what they lacked in range they more than made up for by their ability to move around the battlefield and apply pressure where it was most needed.

Gustavus' final change was to combine the movement of cavalry and guns so that the actions of one arm covered and protected those of the other. His ability to combine fire and movement in the course of a battle became his strongest card, and was highly successful. Unfortunately, he was killed at Lützen in 1632, and his successors, not having his breadth of vision or tactical sense, allowed artillery to lapse once more into obscurity.

If the tactics were indecisive, at least there was steady progress with the equipment. The gun carriage (two wheels connected by an axle, with 'cheeks' to support the gun by its trunnions, and a trail formed of two rearward-facing arms that rested on the ground to support the weight and resist the firing shock) had come into being in the 15th century, and at the beginning of the 16th century came the first artillery instrument, the gunner's quadrant. By now it was realized that there was a constant relationship between the amount of elevation given to the gun and the range to which the ball flew, given that the powder charge was constant, and the quadrant was the means of measuring this elevation. It consisted of a staff with, at its end, a quadrant marked off in divisions or 'points'. Attached to the end of the staff was a cord with a weight. The staff was inserted into the cannon barrel so that the quadrant faced down and the weighted cord cut across it. All that was now required was to elevate or depress the barrel until the cord cut the scale at the requisite point for the desired range.

**Top: Early 16th century cannon and mortars, showing various approaches to the design of carriages and mountings.**

**Below left: A German cannon of the middle 16th century. The chamber, rammer and powder scoop are marked in numbered sections, suggesting that this engraving originally appeared in a gunnery textbook.**

**Right: Gustavus Adolphus, King of Sweden 1611-1632, master of the art of warfare, who revitalized artillery.**

Gustavus Adolphus of Sweden (1594-1632) gave artillery a considerable boost. He was the first to divide his artillery into two branches, field and siege, classing everything above the 12pr (about 4.5in calibre) as siege equipment, and he also introduced the first lightweight field guns capable of being moved at high speed. These were his 'leather guns', lightweight cast copper barrels bound tightly with leather and rope to reinforce them.

Elevating the gun was originally a primitive business; handspikes – long wooden levers – were inserted between the gun and its carriage and the breech end raised or lowered until the correct point was registered on the gunner's quadrant. Blocks were then placed under the breech to hold the gun at the desired elevation. Soon the wedge was adopted; by forcing this in or out the gunner had some degree of fine control over the elevation. The ultimate sys-tem was discovered in about 1571 when John Skinner, an English gunner, devised the ele-vating screw, which was placed beneath the breech and allowed very fine control.

But while the elevation was relatively pre-cise, whichever system was used, direction was simply a matter of the gunner looking across the top of the gun, since there were no sights as we understand the word today.

In the 18th century the War of the Spanish Succession (1702-13) saw a movement to-wards restoring artillery to a mobile role. Marlborough began to demonstrate that tactic-al success could follow if guns were actually moved in battle to take advantage of opportu-nities presented by a careless enemy. At the Battle of Malplaquet, for example, he ad-vanced his 'grand battery' of 40 guns into the heart of the French line where they wheeled to face outwards and poured a withering short-

**A British cast-iron 24pr howitzer of the early 19th century.**

ENGLISH IRON HOWITZER
A mid 19ᵗʰ Century 24 pdr. howitzer mounted on it's original iron garrison carriage.
XIX-297

range fire into the French cavalry reserve, which was wiped off the order of battle before it had a chance to deploy, and thus the course of the subsequent engagement was settled. But with the best will in the world, the guns of the period were still ponderous enough to get left behind if the infantry moved rapidly, and in several of Marlborough's battles his artillery had little chance to shine.

The next important tactical innovation came from Frederick the Great who, in 1759, formed a brigade of horse artillery. Light 6pr guns were used and all the gunners were mounted on horseback so that the brigade could manoeuvre with and keep up with the cavalry. This arrangement allowed the cavalry to function as the shock arm, with lance and sword and thundering charges, but gave supporting firepower when it was needed.

Gradually the other armies of Europe began to follow Frederick's example and give their artillery more mobility. Several systems were tried out; for example, some followed Frederick's method of mounting all the gunners on horses; others put seats on the limbers and guns for some of the gun detachments; others accompanied the guns with waggons carrying the men. Whichever system was tried, the problem was soon seen as a matter of ensuring that gun, gunners and ammunition all arrived at the same place at the same time. The universal adoption of the limber followed from this appreciation. The limber was a light waggon carrying the 'ready-use' ammunition, which was hooked to the gun so that gun and limber and the necessary men were all drawn by the horse team.

The size of the team really settled the size of the gun, and this was to remain the governing factor in gun design until the early years of the 20th century. Six horses were the accepted convenient size of team, which could work well together and be easily managed. The weight was settled at 30cwt (1,525kg) as the upper limit for horse artillery gun which could not gallop into action with more than that. Siege artillery could have heavier guns, since it was appreciated that they moved more slowly and could use larger teams, but even these weapons had an upper size limit.

By the time of the Napoleonic Wars the limber system was universal and all field and horse artillery was capable of quite rapid movement and tactical deployment. By this time, too, there had been some more technical advances.

In the first place, means had been discovered of measuring the performance of the gun

**A contemporary engraving of the Battle of Malplaquet, with the artillery appearing to be in difficulties.**

with some degree of accuracy. Until the 18th century gunnery was more of an art than a science; the gunner knew the idiosyncrasies of his own gun and knew how much powder to use and what elevation to give in order to hit a specific target. But he had little or no idea of the speed at which the cannon-ball moved or what force it carried. In about 1740 Benjamin Robins, an English mathematician, invented the 'ballistic pendulum', a heavy pendulum suspended in a robust frame and with a solid target acting as the weight. A cannon ball was fired at the target; it struck, and thus caused the

pendulum to swing. It was possible by measuring the amount of swing, given that the weight of the target and length of pendulum arm were known, to calculate the force with which the ball had struck. And as the weight of the ball was known, and as the force was the product of weight and velocity, it became possible to calculate the velocity of the ball. Therefore, it became feasible to measure the effect of different formulations and charges of powder and determine which was the most effective, and so ammunition underwent some revision as well.

There were five quite distinct classes of land service artillery by this time. Field and Horse we have already discussed, and they were the immediate support for the manoeuvring army in the field. The Peninsular War brought mountain artillery into use, light pieces that could be dismantled and carried on pack mules. Siege artillery had been in use, of course, for as long as artillery had existed. By this time it mustered two distinct types of weapon, the mortar and the howitzer. The mortar had been invented by the Turks; in 1451, at the siege of Constantinople, Mohammed II was confronted by an enemy fleet anchored off the Golden Horn and 'proposed... a different mode of proceeding and a totally new description of gun, of which the form should be a little modified so as to enable it to throw its shot to a great height that in falling it might strike the vessel in the middle and sink her...' The mortar was duly constructed and sank a warship with its second stone ball. Thereafter the idea of using a high trajectory weapon rapidly spread through Europe. Strangely, the original Turkish intent, that of sinking ships, was forgotten – probably because it called for a degree of accuracy that was rarely found in mortars – and the mortar was used principally as a siege weapon since it could loft its projectile over walls and defences.

**Above: A 16th century engraving which illustrates the techniques involved in demolishing towers: either by concentrating fire on one spot or by cutting across the target.**

The mortar was a short-barrelled weapon fired from a 'bed', a simple slab of wood; it had little or no adjustment for elevation, the range being controlled by adjusting the powder charge. The howitzer, on the other hand, while also intended for high trajectory shooting, was more like a conventional gun. It was fired from a two-wheeled carriage and was capable of being adjusted in elevation, though some degree of control was also given by altering the powder charge. But the limitations of the gun carriages of the day meant that the howitzer's maximum elevation rarely exceeded about 20°, and thus it was not strictly comparable with the mortar. Where it did differ significantly from the gun was in its projectile; whereas the gun primarily fired a solid ball, the howitzer was usually provided with a hollow ball containing gunpowder, which would explode and cause damage and fires within the besieged town.

The last category of artillery was the Garrison Artillery, weapons used to defend a fort or to oppose a sea landing from a permanent defensive work. The weapons were more or less the same guns as those used by the field

**The 17.72in rifle muzzle-loading gun of 'Napier of Magdala' battery at Gibraltar.**

and siege gunners, but were usually of large calibre and mounted on four-wheeled 'garrison carriages', which were excellent for lining the ramparts of a castle but would have been useless for moving with a field force.

When the Napoleonic Wars ended, and Europe entered into the 'Long Peace', artillery had, in fact, changed very little across the centuries, and a reincarnated gunner from the army of Frederick the Great, Cromwell or Marlborough would have found very little in

**Inset: A battery of smoothbore cannon on the ramparts of Plymouth Citadel.**

the guns of the 1830s to puzzle him. But with the Industrial Revolution now underway, engineers and mechanics were inventing new machines and devices every week, and it was only a matter of time before one of these men turned his attention to the then current state of artillery. Incentive was all that was needed, and this appeared in 1854.

In fact, there had been new ideas before that date. In 1821 a Lieutenant Croley of the British Army had proposed a breech-loading rifled gun firing a lead-coated shell; in 1846 Baron Wahrendorff of Sweden made a similar proposal; in 1842 the Chevalier Treuille de Beaulieu of the French Army had designed a muzzle-loading gun with a few deep grooves and a shot with lugs to ride in the grooves. Some of these proposals were made and tested, but none appeared to be sufficiently reliable nor sufficient improvement over the existing cannons to make its full-scale manufacture worthwhile. What had prompted all these proposals was simply the threat of rifled firearms in the hands of infantry; an infantryman with a rifle-musket was capable of shooting a gun's detachment of men at ranges at which he was virtually safe from retaliation, except from a very lucky shot. The range and accuracy of small arms were becoming a serious threat, and artillery, unless it could be improved to a similar degree, looked like being chased from the battlefield.

The Crimean War broke out in 1854, and a Mr. William Armstrong, an English lawyer turned engineer, was astonished to read in his newspaper that British 18pr gun weighed over three tons and was less than mobile in the mud of the battlefield. After studying the current method of gun construction he decided that something better could be produced by using more scientific methods. He sat down and designed the first 'modern' piece of artillery.

Armstrong's system was to build up the gun from a number of tubes or 'hoops' of wrought iron, instead of casting the entire barrel out of one lump of iron. He prepared the rifled barrel, then carefully dimensioned another tube to fit over the barrel; this was fitted by heating it so that it expanded, to be slipped over the barrel and allowed to cool. As it contracted, the outer hoop gave strength to the inner barrel and resisted the internal explosion of the charge. More hoops were shrunk on, the thickness being graduated so that there were more hoops over the gun chamber, where the pressure was greatest, than over the 'chase' (that part of the barrel in front of the trunnions).

The Armstrong gun was also a breech-loader; the rear end of the barrel had a vertical slot, and the portion of the tube behind this slot was screw-threaded. A wrought iron block was dropped into the slot to close the rear of the chamber, and a heavy screw, working in the rear section of the barrel, could be turned to jam the block securely against the barrel and prevent any leak of gas. Sealing was assured by using a copper facing on the block and carefully machining the mouth of the chamber. To load, the block was removed and a shell rammed into the chamber through a hole in the locking screw; the shell was lead-coated so that the lead bit into the rifling grooves in the barrel. The cartridge – a cloth bag of powder – was inserted in the chamber behind the shell, the block dropped into the slot, and the screw tightened up. Firing was effected by inserting a 'friction tube' into a vent drilled through the block. The friction tube was a quill containing fine gunpowder and with a form of friction match composition at the top, which was ignited by pulling a rough metal bar through the composition, which fired the fine powder, flashed down the vent and ignited the powder.

**Left: The boredom of war: a mortar crew at rest outside Sevastopol during the Crimean War, 1854.**

**A contemporary print of the Siege of Sevastopol, indicating rather more activity and excitement than the photographer found.**

Armstrong delivered a 3pr gun to the British Army for testing in 1855. The Army was impressed, but it was thought that to adopt the system without offering other systems a chance would be unfair. The Armstrong Gun emerged victorious from the series of trials that took place with other designs, and in 1858 it was formally recommended for adoption. In the following year Armstrong assigned his patents to the Crown in return for which he was appointed 'Engineer for Rifled Ordnance to the War Department' and Superintendent of the Royal Gun Factory at Woolwich Arsenal. At the same time he retained control of his own factory in Newcastle, which became the Elswick Ordnance Company and manufactured Armstrong guns first for the British Army and then for export all over the world.

As ever, the armies of Europe were looking over their shoulders and watching what their foreign counterparts were doing, and the Brit-ish Army's wholehearted adoption of Armstrong's system led to action in France and Germany. The French adopted de Beaulieu's system of rifling. These bronze smoothbore guns had six deep grooves cut into their bores, and the projectile had two rows of soft metal studs let into its surface. The charge having been loaded, the shell was then introduced into the muzzle base first, the studs being carefully fitted into the rifling grooves. As the shell was rammed down to the chamber, so the studs rode in the groove and rotated the shell; in a similar way, when the charge exploded and blew the shell out, so the groove imparted the desired rotation by means of the studs.

The Prussian Army went even further than merely adopting rifling and breech-loading and began to adopt steel as a material for gun construction. This may seem an obvious step, but in the 1860s steel was a somewhat problematic material and there appeared to be great difficulties in manufacturing a piece of steel that had the same quality all the way through. Gunners distrusted steel, because experimental guns of steel on trial had shown that when they eventually blew up they tended to do so violently and with dangerous results. A faulty wrought iron gun, on the other hand, would give way gradually, splitting first and thus giving the gunners sufficient warning for them to avoid any serious injury.

But Alfried Krupp, the steelmaster of the Ruhr, was rapidly becoming the leading expert on the manufacture of high-grade steel. After years of trials, he produced a steel 90mm gun in 1856 which the Prussian Army adopted and found so satisfactory in service that henceforth all its guns were built of Krupp steel. Krupp also developed a unique breech closing system, using a solid wedge of steel that slid sideways in a mortise in the rear end of the gun. By inclining the mortise guides, the block was

A British squad at drill, c.1865, about to lower a mortar to the ground and remove its wheels, preparatory to firing.

thrust tightly against the chamber as it closed, and a soft metal ring was let into the face of the breech – a system often called the 'Broadwell ring' after its originator – and clamped tightly against the end of the chamber to make an effective gas seal. Krupp used deep rifling and studded shells in his first designs. Initially these were similar to those found in the French system, but he was not satisfied and changed to fine rifling and lead-coated shells after Armstrong's design, and eventually settled for fine rifling and a soft metal cup on the bottom of the shell which bit into the rifling.

Although the prime reason for rifling was to increase the gun's range, there were other advantages. In the first place the size of the projectile could be increased. A round ball to fit a given calibre was a finite size and governed the weight of projectile; but an elongated shell of the same calibre could be much heavier and thus more effective at the target. But an elongated shell could only be fired in a rifled gun that gave it stability by spinning it; fired from a smoothbore, it would simply turn end-over-end and be hopelessly inaccurate. Rifling also improved accuracy by giving the shell stability, and it also improved range because an elongated shell with a pointed nose offered less resistance to air than did a round ball.

Below: A Prussian siege battery outside Strasbourg during the War of 1870. The howitzer will recoil and ride up the iron ramps placed behind the wheels, a system that was still used into the World War Two.

**A 12.5in rifled muzzle-loading gun on its original carriage, now a roadside ornament at Gibraltar.**

It should be said that there were other systems of rifling that did not involve having grooves in the gun's bore. Foremost among these were the systems of Lancaster and Whitworth, both British engineers. Lancaster used a gun with an oval rather than round bore and cut the oval so that it made a gradual twist from chamber to muzzle. The projectile was an elongated slug of metal cast in a twisted oval section to match the bore. When fired it rode on the twisted oval and developed spin. Whitworth's system was very similar, except that he used a bore of twisted hexagon shape and a shot with the edges planed into a similar warped hexagon. Both were effective when new and when the ammunition was carefully manufactured and carefully loaded. But once the guns began to wear, and when the ammunition was mass-produced with rather less care, problems became apparent. The principal one was that the shot frequently wedged itself in the bore, which placed the surfaces of shot and gun out of alignment with each other. If this happened during loading it was a considerable nuisance; if it happened when firing it could be highly dangerous, leading to a burst gun. Both systems were tried, but neither was adopted in any quantity, even though a number of Whitworth guns was supplied to the Confederate Army during the American Civil War.

Meanwhile the British had been using their Armstrong guns in combat and discovering a few drawbacks. The Army in the New Zealand Wars, the Navy in various fleet actions around Japan and China: each complained that the breechblock – or 'vent piece' as it was then known – was prone to blow out of the gun from time to time. This problem was generally considered to be the gunners' fault for failing to operate the locking screw tightly enough. But the copper facing was also giving trouble and there were dangerous leaks of gas past the front of the vent piece. Various modifications and improvements were made, and the Armstrong Gun remained in field service; but its utility for the Royal Navy had suddenly been put into considerable doubt by developments in France.

In 1858 the French Navy laid down a new type of warship; its *La Gloire* was a wooden ship clad with a layer of thick iron armour, and overnight it made the wooden ships of the rest of the world obsolete. The Royal Navy retaliated by building a totally iron ship, *Warrior*, and at the same time began some gunnery trials to find out how to defeat iron armour. The Army became involved at this point, because it would require coast defence guns capable of dealing with ironclad warships.

**Above right: A typical siege gun of the 1860s in position, protected by field fortifications.**

**Right: The 2nd Surrey Artillery Volunteers at practice on the Plumstead Marshes, near London, in 1865.**

It soon became apparent that the Armstrong gun was almost useless in this new role. The breech closing system was simply not strong enough to withstand the explosion of large charges of powder to send heavy shot with sufficient velocity to smash through armour; the vent piece invariably blew out. The only way of achieving the immense power required to propel a heavy shot with sufficient force to go through several inches of wrought iron, plus several more inches of the teak-wood backing, was to use a muzzle-loading gun with a heavy, solid breech end that could not be deranged in any way by heavy charges. At the same time, though, it was appreciated that the advantages of rifling were too good to lose, and Armstrong promptly developed a muzzle-loading gun with a system of rifling based on

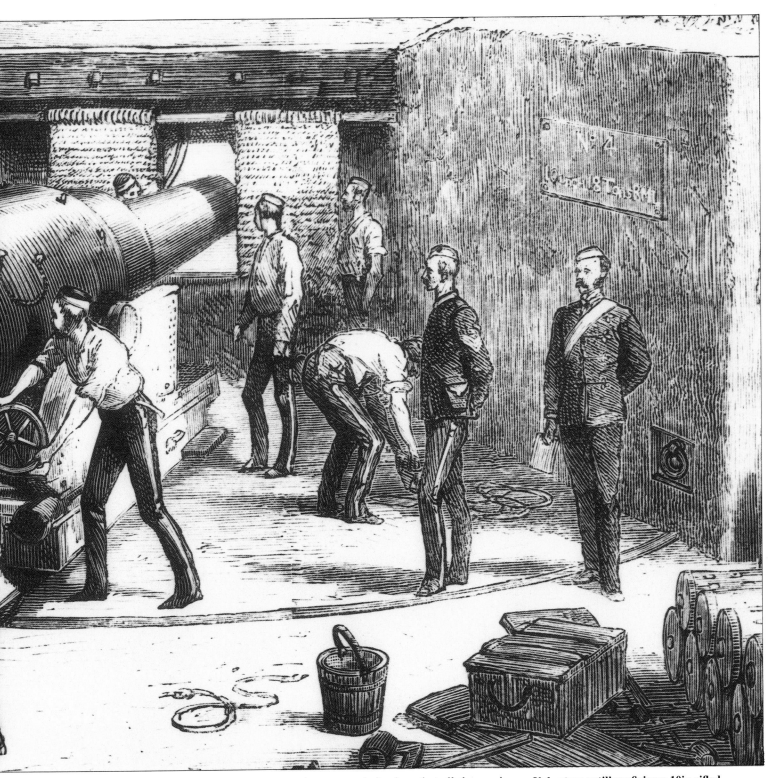

that of de Beaulieu, three deep grooves firing a projectile with two rows of soft metal studs that engaged in the grooves. When put to the test it was shown to be superior to the rifled Armstrong gun in destructive power; it was also a lot simpler to use and maintain in action and, as a result (and so that all the services would use a similar system), the Armstrong gun was superseded by the 'RML' or Rifled Muzzle Loading gun in the mid 1860s. The change from breech-loading to muzzle-loading is often derided as a typical reactionary

**Volunteer artillery firing a 10in rifled muzzle-loading gun at the Shoeburyness School of Gunnery, about 1880. The emplacement is representative of a coast fortress, an iron shield between two granite wings.**

**A contemporary photograph of the British 9in rifled muzzle-loading coast defence howitzer.**

step by the highly conservative military establishment of the day, but that charge is a gross over-simplification. It was forced on the military by the combination of a peculiar tactical demand and the inability of the engineers of the day to produce a sufficiently strong breech mechanism.

However, in the reversion to muzzle-loading, the advantages of Armstrong's method of building up the gun from successive hoops of metal were not abandoned, although the design was simplified; and instead of half-a-dozen hoops, there were but three or four in the heavy RML guns. The threat of the ironclad also led to an orgy of fort-building all over the Empire, and these forts had to be armed. The new RML guns were expensive, and there was a number of schemes put forward for rifling old smoothbore cannon so that they would be able to function as RMLs. Most of these simply cut grooves into the existing metal, and inevitably the gun blew up after a few shots; the original design was not strong enough, and carving the grooves further weakened the

barrels. The correct solution was found by Captain Palliser, a cavalryman, who developed a rifled wrought-iron tube that could be inserted into an old smoothbore, expanded into a perfect fit by firing a charge of gunpowder inside it, and then vented. The result added strength to the original gun, and hundreds of 'RML Converted' guns were turned out in order to provide cheap armament for the growing number of forts and coast defence batteries.

Although the shape and construction of guns were undergoing changes, the one thing that had remained the same was the propellant: gunpowder was the only explosive that could be used to shoot projectiles out of guns. It had its drawbacks, as it had always had; dense white smoke, fouling, susceptibility to damp, liability to accidental explosion from friction, all these were well known and had to be accepted as the inevitable facts of military life. But hand-in-hand with the revolution in engineering came advances in chemistry, and explosive chemistry was being thoroughly

investigated. The first advance came with the invention of guncotton in 1846; this was cotton that had been acted on by nitric and sulphuric acids, and it proved to be a powerful explosive. Manufacture began in Britain, France and other countries in 1847, but it was soon apparent that there was something unpleasant about guncotton; it had a tendency to explode for no apparent reason, and after some disastrous accidents in factories and stores,

most countries banned its manufacture. It was not until the 1860s that experimenters proved that the trouble lay in insufficient purification of the finished product, leaving acid residues that eventually led to spontaneous combustion. Guncotton became a perfectly safe explosive once this was cleared up. However, in spite of many trials, it was far too violent for use as a gun propellant, and gunpowder remained the standard.

In order to improve its performance, gunpowder was now being made in various sizes and shapes by compressing the damp powder into blocks, prisms, pellets or hexagons; every experimenter had his own ideas on the subject. The object was to slow down the explosion so that the shot was given a sustained push rather than a violent acceleration. In order to get the best results from this type of powder, the gun needed to be long to give the slow-burning

**A 32pr muzzle-loading cannon in an embrasure at Tilbury Fort.**

powder sufficient time to complete its combustion and generate all the gas possible before the shell left the muzzle. This requirement conflicted with the need to keep a muzzle-loading gun short so that the gunners could get round to load it without having to pull it back inside the ship or fort; so breech-loading had to be perfected.

29

In 1878 this had become apparent to the British Army, which began to cast about for a reliable breech-loader. It will be no surprise to hear that who should be waiting for the call but Sir William Armstrong (as he had now become), who submitted a 6in and an 8in gun for trial. These used a system of closing the breech which had originated in France: the interrupted screw. The rear end of the gun chamber was screw-threaded, and slots were then milled out of the threads. A heavy steel cylinder was also screw-threaded to match, and it, too, had slots milled across the threads to result in a piece that had a series of threaded sectors. The screw could now be inserted into the breech of the gun by sliding the threaded sectors into the milled-out sectors of the chamber. Once fully inserted, the screw was turned so that the threaded portions engaged with the threaded sectors inside the chamber, and securely locked the breech block in place. The strength of the lock was ensured simply by having sufficient threads in engagement; the bigger the gun, the longer the breechblock and the more the threads.

This locked the block in place but obviously did nothing about sealing the explosion gases inside the breech, because without some special arrangement they would simply leak out through the slots. Armstrong's solution was to place a metal cup on the front end of the block. The edges of this cup slid snugly against the wall of the gun chamber, and when the charge fired these edges were forced out by the pressure to make a seal. The idea worked, Armstrong's 6in gun was a success, and breech-loading came back into use and replaced the rifled muzzle-loaders, though many of the RML guns actually stayed in use until the end of the century.

**Inset: British field artillery with 12pr breech- loading gun, during the New Forest Manoeuvres of 1895.**

**A 6in breech-loading gun of Devil's Gap Battery, Gibraltar, typical of the coast defence guns used until the 1950s.**

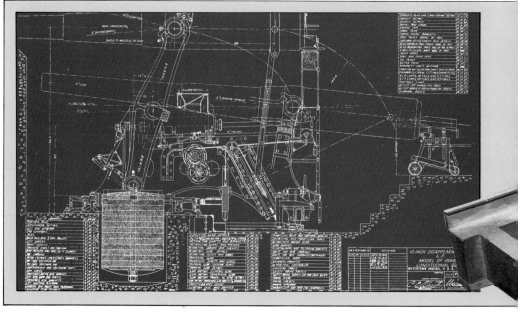

A Krupp breech-loading gun captured from the Boers during the South African War and now outside the Royal Artillery Museum, Woolwich.

Armstrong's 'tin cup' sealing system was not entirely successful; like many other ideas it worked well when it was new and was being demonstrated by the manufacturer's skilled men, but once in service, with averagely careless soldiers, and after some amount of use, the cup frequently failed to make a satisfactory seal. The same problem had been faced elsewhere and, once more, the answer came from France where an officer named de Bange had developed a breech sealing system for use with the interrupted screw.

De Bange's breech was in two parts, the breechblock itself with the interrupted threads and with a hole through its centre; and the 'mushroom head' or vent piece, given its nickname from its shape, a long stem that passed through the hole in the screw and had a rounded head of the same diameter as the interior of the chamber. Between the two was a ring-shaped pad of resilient material; the first designs used a compound of asbestos fibres, oil and brass wire mesh. With the breech closed, the mushroom head and the sealing pad lay inside the chamber. When the charge exploded the pressure drove the mushroom head back and so squeezed the pad between the mushroom and the front end of the breech screw. This caused the pad to expand outwards and press tightly against the wall of the chamber to effectively seal the rear of the gun against the escape of gas. With improvements in the nature of the material used for the pad, the de Bange system is universally employed today as the principal method of sealing the breech of guns when bagged charges are in use.

The only serious competitor to the de Bange system was that of Krupp. His sliding block had begun by using the Broadwell ring with bagged charges, but in the 1870s Krupp began experimenting with brass cases for the charge. These were essentially enlarged versions of the brass case adopted for use with small arms, and were designed so that the elasticity of the brass allowed the case to expand against the chamber to make a seal and then contract slightly when the pressure dropped to allow the case to be extracted. Well suited to his sliding block, Krupp's system became the German standard, and was eventually used with guns as big as 80cm calibre.

The next thing to attract the attention of inventors was the problem of dealing with the recoil of the gun when it fired. Recoil had been a problem from the earliest days of artillery; when a gun is fired the reaction to the expulsion of the shot causes the barrel to move backwards and, of course, take the carriage with it. The only remedy was to allow the gun to recoil and then laboriously push it back to its firing position. This was considered acceptable practice with light field artillery and shipboard guns, even though, at the end of a prolonged battle, artillerymen were often firing the guns from where they had come to rest, having exhausted their strength for pushing them back.

The amount of recoil depends upon the relative sizes of the shot and the gun, and when guns and their projectiles began increasing in size the recoil problem suddenly became serious. The first attempts were made with heavy coast defence and fortress guns, and consisted simply of fitting the gun into a carriage and then placing the carriage on top of a 'slide', a wooden framework that was inclined a few degrees upwards to the rear. On firing, the wheelless carriage slid back on top of the slide so that friction and the upward angle gradually slowed the movement and brought the gun and carriage to rest. If heavier charges were fired, a handful or two of sand flung on the slide would add friction or, if reduced charges were fired, a layer of grease would reduce it. The object, in both cases, was for recoil to stop in a position convenient for reloading the gun. The whole carriage and gun had then to be manhandled down the slide again to the firing position. The addition of wheels on eccentric axles helped; they could be thrown out of contact for the recoil movement, then brought into play to aid in returning the gun.

Next came the 'compressor', an arrange-

**Above: A wooden model of the Moncrieff disappearing mounting, with a 7in rifled muzzle-loading gun.**

**Far left: Showing the complexity of fortress artillery, the American 10in disappearing gun carriage Model 1894.**

ment of iron plates suspended beneath the gun carriage and moving between a pack of similar plates held on the slide. The plates could be squeezed together by turning a screw-press to provide the necessary friction to damp out recoil, and they could be loosened to allow the gun to be returned 'into battery'.

Recoil is wasted energy, and thus the next step was remarkable in that the inventor actually put the recoil to some useful work. Captain Moncrieff of the Edinburgh Militia Artillery was mainly concerned with concealing the gun from the enemy between shots, and his 'disappearing carriage', developed in 1865, being intended for emplacement in a pit for use in coast defences. The gun was attached to the top of two curved 'elevator arms' that had a heavy mass of iron and lead at

the foot. With the arms vertical, the gun pointed over the parapet of the pit. When it was fired the recoil drove it backwards, and the curved arms rolled back on their slide, lifting the counterweight and lowering the gun into the pit. Once down it was stopped by a friction brake and held down while it was reloaded. Once ready, the brake was released and the counterweight now pulled down the front ends of the arms and lifted the gun above the parapet for the next shot.

The disappearing carriage was more expensive than the common carriage and slide, but its advantage of concealment between shots meant that the emplacement was much cheaper than the usual system of building powerful granite and armour forts, and the system was embraced with enthusiasm. But it was soon

discovered that, for various reasons, it was not suited to guns of more than about 7in calibre and, because the size of guns was now going well beyond that, something more was needed.

The next step was the adoption of hydraulic brakes to control recoil. In their first, simple form these were cylinders attached to the slide with a piston and rod attached to the gun carriage. The cylinder was filled with oil, and the piston had a number of holes in it. On firing, the gun and carriage began moving back along the slide, pulling the piston through the oil. Oil being incompressible, it had to escape through the holes in order to pass from one side of the piston to the other, a process which absorbed the recoil energy and controlled the movement. The gun had to be

pushed back after loading and the oil then passed through the holes once more.

The hydraulic principle was then applied to the disappearing carriage; the 'gun arms' that held the gun were joined by a transom, which carried a piston rod connected to an oil cylinder in the mounting. On firing the arms pivoted back, pressing the piston into the cylinder, and the oil was forced out through a restricting valve into another cylinder containing compressed air. The injection of oil further compressed the air which acted as a cushion and gradually brought the gun to rest in the 'down' position, ready for loading. As it came to rest, the valves automatically closed, keeping the pressurized oil and air locked into their cylinder. To lift the gun, the valves were manually opened and the compressed air forced the oil back into the mounting cylinder and pushed on the piston, so lifting the gun back into the firing position. Disappearing carriages on this pattern were developed for guns up to 10in calibre, and in other countries, notably the USA, modified versions of this carriage were used with guns up to 16in calibre.

The hydraulic buffer was now adapted to control recoil in a number of fixed mountings for coast defence and naval use, but it was not until 1897 that the first successful hydropneumatic recoil system appeared on a field artillery weapon. This was the 'French 75', or the '75mm Canon Modèle 1897', and it revolutionized the design of field artillery. For the first time a field army had a gun with efficient recoil control, which meant that the gunners, instead of standing clear on firing to allow the gun to recoil, could group tightly around it. This, in turn, meant that it was feasible to protect them by fitting a shield to the gun. And, as the men were there, they could be ready to load as soon as the gun had fired, so it was given a quick-acting breech mechanism and a fixed round of ammunition; that is, a round in which the brass cartridge and the shell were fixed together and could be loaded in one movement. The result was that the French 75 could generate a rate of fire of 20 rounds a minute, which was unheard of in field guns. It suited the French tactical thinking to have a blast of fire to accompany the infantry in the

**Above: British 12pr guns in action at the Modder River, South Africa, 1899.**

**Right: British 5in siege gun firing, South Africa, 1900.**

assault and, of course, it rendered every other field gun in the world obsolete overnight.

The 1890s also saw the widespread adoption of small-calibre high-velocity magazine rifles for infantry and, as had happened 60 years before, the fire of infantry threatened to drive artillery from the field. This superiority became apparent in the South African War when British field artillery was deploying out in the open alongside its infantry – as it had done for centuries – and being shot to pieces by Boers firing at long ranges from concealed positions. The artillery had to conceal itself in order to survive, which posed the problem of how to shoot at targets from concealed positions. Before this, all artillery shooting had been 'direct fire', with the gun and its target in sight of each other.

**Above: A Boer 75mm field gun of French origin, behind a stone breastwork overlooking the Tugela River during the Battle of Colenso, December 1899.**

The solution lay in the development of the goniometric sight, later to be called the 'panoramic' or 'dial' sight. This instrument was fitted to the gun mounting and could be pointed in any direction, which could be accurately set by a scale of degrees. If the target and the gun's position were now located on a map, the angle between the two could be determined. The gunner now looked for some prominent object visible from his gun position – which was, remember, concealed from the enemy. The object could be identified on the map and the bearing to it calculated. If the goniometric sight was now set at the angle between target and aiming mark, and the gun swung until the sight was laid on the mark, then the gun itself would be pointed at the target. Not with great precision, is it true, but with sufficient accuracy to place the first short somewhere close to the target. The next move was to send an officer forward until he could see the target and spot the fall of shot. He could then send back orders to the gun to move right or left, add or drop range, until the shots fell in the right place.

Most countries were adopting methods of on-carriage recoil control at the beginning of the 20th century, which frequently consisted of a combination of oil buffer and a bank of springs to return the gun to battery once recoil had ceased. At the same time the techniques of indirect fire were being explored and perfected. The conventional field gun of the day was of about 75mm calibre and designed to fire a shrapnel shell against troops in the open. It was generally augmented by a howitzer of about 100mm calibre, firing a powder-filled shell at a steep angle to pass over defences and act as a destructive weapon against matériel. By this time smokeless powder had been developed to replace gunpowder as a propelling charge, and first trials of high explosives for filling shells were underway. The near-universal choice of explosive was picric acid, melted and cast into the shell; this was sufficiently stable to withstand the shock of firing but was also sufficiently sensitive to be detonated by a fuze when it struck the target. The only drawback was that the acid reacted with the metal in the shell and developed highly sensitive salts that could detonate if the shell were dropped, but this was overcome by varnishing the interior of the shell.

The next major step in the progress of artillery came with the outbreak of war in 1914. The German plan of attack called for a rapid advance across Belgium, and for the destruction of some very strong chains of forts that the Belgians had built across their path. These forts, built in the 1880s, had been designed to withstand the heaviest artillery that a field army could bring to bear. The figure arrived at was governed, as always, by the weight that could be pulled by horses and was reckoned at 24cm calibre. The forts therefore had sufficient thickness of concrete and iron armour to stop a 24cm shell.

To everyone's surprise, the German Army arrived outside Liége with two enormous howitzers of 42cm calibre, and within three days had thoroughly wrecked most of the forts

**Inset: Five tons of 60pr gun, deep in the Flanders mud, 1915.**

**Right: A German 15cm howitzer used by the Portuguese Army until the 1970s.**

**Below: A gun cupola of Fort Loncin, Liége, after being shelled by the German 42cm howitzer 'Big Bertha', 1914.**

and captured the Liége fortress. The performance was repeated at Maubeuge and other fortresses when it became obvious that the old rules no longer applied.

The 42cm weapons had been designed by Krupp as coast defence howitzers; he had modified the design so that they could be dismantled by crane and moved piecemeal on railway trucks. When offered to the German Army, the idea was considered but passed back to Krupp with the request that he make the weapon movable by road, because there were now motor tractors capable of towing heavy loads. Krupp had redesigned it again, so that it moved in five loads each hauled by a tractor, and could be rapidly assembled by using a portable hoist. The result was the famous 'Big Bertha', of which two were built and which did most of the damage at Liége.

It would be pointless to detail every gun designed and built during the 1914-18 war because it was essentially an artillery war, and types and modifications proliferated. But it was this war that produced the anti-aircraft gun in quantity, the railway gun, the anti-tank gun and a variety of long-range guns, each of which was a response to some particular tactical challenge.

Anti-aircraft guns had first appeared in 1909, as 'anti-balloon' guns; they were often mounted on truck chassis because the slow speed of airships and balloons encouraged the idea that mobile guns would be able to chase their targets, but this was soon found unpractical. There was very little difficulty in designing guns to shoot into the sky once the requirement existed; the problem was in hitting what they were shooting at. The problems of firing at a target that was moving fast and capable of moving in three dimensions were enormous, and some complicated sighting apparatus was put on the guns. Eventually it was realized that a more practical solution was to put the 'sight' in the centre of the battery and pass the necessary elevation and direction to the guns rather than put an expensive sight on each weapon and surround it with operators.

**British 8in howitzers of the 39th Siege Battery, in action during the Battle of the Somme, 1916.**

**Above: This 3in Model of 1918 was the first American anti-aircraft gun for use with the field armies.**

**Right: A French 75mm gun Mle 1897 removed from its carriage and fitted to an improvised mounting to turn it into an anti-aircraft gun for the defence of London in 1915.**

**Left: American 8in howitzer Mark 6 on self-propelled Holt carriage, 1918; one of the first self-propelled weapons.**

Anti-tank guns did not appear until the closing days of the war, when the Germans produced a very low-set 37mm gun intended for concealment until the tank came close before opening fire. With the thin 'armour' of the tanks of the day it would have doubtless been lethal, but in fact none appears to have been used.

Railway guns were not new; they were first tried by the Americans during the Civil War, and French companies had made several in pre-1914 days for sale to South American countries as mobile coast defence weapons. But the peculiar conditions of the Western Front, with its ample networks of railway lines close behind both trench systems, meant that heavy railway artillery could be moved around rapidly to confuse the enemy and also to add support to different sectors at different times. These were invariably big guns; there was no point in putting a field piece on a railway mounting, and so guns and howitzers of 8in to 20in calibre were placed on railway trucks. Most of these had to be specially designed to take such huge weapons and stay within the loading gauge, and various sytems of controlling the recoil were adopted. Some carriages

were lowered from their wheels to rest on the track, and slid backwards a few feet when fired; others were allowed to roll back on their wheels and were then returned to position by a locomotive; a few used hydraulic recoil systems to soak up most of the recoil before it began to move the carriage. But since the object was generally to get a heavy gun into action cheaply, the introduction of a recoil control between the gun and the mounting was often thought to be an unnecessary expense, and simple sliding mountings were popular.

Long-range guns were developed in order to upset rear areas with harassing fire; headquarters, supply dumps, railway junctions could all be given sleepless nights by a few shells flung over at odd intervals, which could be done with impunity by a long-range gun because the enemy would not have anything powerful enough to retaliate. Most of these were naval guns of 12-14in calibre on railway mounts, but there were one or two exceptions; the most famous was Kaiser Wilhelm Geschütz or, to give its more common name, the Paris Gun.

**A Krupp 1000pr siege gun, used at the siege of Paris in 1870.**

The Paris Gun was the creation of Professor Rausenberger, Krupp's ballistic expert, a man who had already amused himself by designing a gun to shoot across the Alps. It was a 38cm naval gun with a liner inserted in the barrel to reduce the calibre to 21cm. Ordinary rifling would not suffice, so he rifled the gun with a few deep grooves and fitted the shells with curved ribs that would ride in the grooves to give the necessary spin. The gun launched its shell into the stratosphere with the help of an enormous charge. Air resistance is negligible at that altitude, and thus the shell coasted a long distance before it re-entered the earth's atmosphere. The guns were emplaced north of Paris and opened fire at Easter 1918 with a range of 76 miles; their accuracy was not

good, but with a target the size of Paris it was impossible to miss, and the Parisians were severely shocked to find they were being attacked by a gun. The enormous charges soon wore the guns out, forcing the Germans to bore the barrels out to 24cm, to permit firing to continue. Despite intensive aerial reconnaiss-ance the Allies never did discover their posi-tion. Only when the advances in the 1918 offensives caused the guns to be moved did the threat vanish; so did the guns—they were never seen again and were presumably scrapped before the Allied Disarmament Commission could find them after the war.

**Above: Moving a British 60pr gun forward for the Battle of the Scarpe, near Tilloy, April 1917.**

**Below: A British battery of 18pr field guns, France 1916. By this stage of the war deployment in the open, as seen here, was the exception.**

Above: A British twin-barrelled 6pr coast defence gun, used with great effect on several occasions against torpedo-boats and raiders.

Right, above: This American 14in gun was developed as a mobile coast defence weapon; two were emplaced in the Panama Canal Zone and two others on the West Coast of the USA in the late 1920s.

Far right: The first British mobile anti-aircraft gun was the 3in mounted on a 'Peerless' truck. Introduced in 1917, it remained in use until 1940.

Left: 242nd Coast Artillery Regiment (Connecticut National Guard) firing a 12in gun during its annual practice at Fort H.G. Wright, New York, in the 1920s.

As ranges of guns increased, so did the problem of hitting the target. In the case of the Paris Gun, for example, it was obviously unpractical to hope to send a German officer through the French lines to spot the fall of shot in Paris. In fact, the Germans relied upon spies and a line of communication through Switzerland for their information, but it could scarcely be called 'correction of fire'. As a result, the use of 'predicted fire' began to be explored. By this time there was some considerable knowledge about what happened to the shell in flight, how it was acted upon by winds, air density, temperature and other phenomena. Attempts were made to forecast these effects and make corrections to the gun's direction to try and compensate and ensure that the shell landed where it was intended. The artillery had to be given accurate weather information and had to know very precisely how its guns were affected by the different factors. Predicted fire began in 1917-18, but it was to be many years before it was perfected.

The period between the two world wars saw

a great deal of research but very little change in artillery until the mid 1930s. The feeling in the 1920s was that another war was unlikely, and very little money was available to armies for equipment; moreover, most had all the equipment they needed, left over from the war. Work went instead into the development of anti-aircraft guns and into the predictors that would calculate the firing data against a moving target, and on the development of a suitable counter to the tank. Nations with special military problems developed special solutions; thus Britain, faced with arming the great naval base of Singapore, developed a 15in coast defence gun and also a twin-barrelled 6pr anti-torpedo-boat gun with a very high rate of fire.

The slumber ended when Germany began to re-arm, from about 1934 onward. This stimulated the other nations of Europe, and artillery was high on the list of priorities, because all the existing weapons were of World War One or earlier vintage. One of the most vital tasks was to adapt guns to mechanical traction, either by substitution of pneumatic-tyred wheels or by fresh designs; another was the provision of air defence against the bomber which was the universal bogey of the times.

Germany, having had most of its armaments scrapped or confiscated after 1918, was able to start with a clean sheet and produce a host of modern designs. Other countries, less dedicated to the military ethos, were slower in producing, even if they had good designs prepared. America, for example, had spent most of the 1920s developing sound designs for an entire re-stocking of its artillery, but

financial restraints meant that very little was actually made. But when war loomed, in 1940, the Americans were able to reach into this stock of designs and put weapons into mass production with little delay.

Of the artillery of the Second World War, a few designs stand out, either because of their technical features or because of their tactical ability. One of the most prominent of the latter

was the German 'Eighty-Eight', the 8.8cm FlaK18 anti-aircraft gun, designed by Krupp technicians working with the Bofors Company in Sweden in the late 1920s. When Hitler took power and the climate of opinion seemed right, Krupp brought these designers back to Essen where they perfected their gun design. It was built, approved, and went into mass-production in 1934. There were periodical improvements to it as new ideas occurred. Some went to Spain with the Condor Legion where their possibilities as anti-tank weapons were first explored and as a result, special anti-armour ammunition was designed and manufactured, well before 1939. In 1940, when a British armoured column broke through the German lines at Arras to threaten a Panzer formation, only the hurried deployment of a handful of 88s saved the German force. It became the scourge of British armour in the Western Desert, and led to the development of 'proper' 88mm anti-tank guns which

would save the Germans once more, in Russia – at least, for a time. But the point about this is not that the 88 was some technical freak, capable of incredible feats of shooting; in fact, it was inferior in all respects to the British equivalent, the 3.7in anti-aircraft gun: in range, ceiling, rate of fire and shell weight. Where the 88 scored was in quantity. The German Army was never short of them and commanders could deploy them as anti-tank guns. The British, on the other hand, never had enough 3.7in guns to fill the air defence role, and they certainly could not afford to take them from this task and deploy them as anti-tank weapons. Moreover, nobody had given the idea much thought. The guns did not have sights suitable for the anti-tank role, nor were they provided with anti-armour ammunition. And thus, by default, the 88 has passed into legend while the 3.7in has merely passed into history.

Much of the technical innovation in the

**Above: A 40mm Bofors anti-aircraft gun, designed in Sweden, manned by troops of the Indian Army in Italy, 1943.**

**Right: A British 3.7in anti-aircraft gun on the outskirts of London, November 1940.**

1939-45 period came from Germany, principally because the Germans had their stock artillery equipments in production, modifying them periodically, and were able to devote time to thinking up new ideas. Britain, Russia and the USA, on the other hand, had their hands full simply producing enough artillery for their day-to-day requirements, and had less opportunity to spare design staff for flights of fancy. This is not to say that the Allies did not develop new ideas; they did, but they were more closely matched to specific operational requirements than were the German innovations.

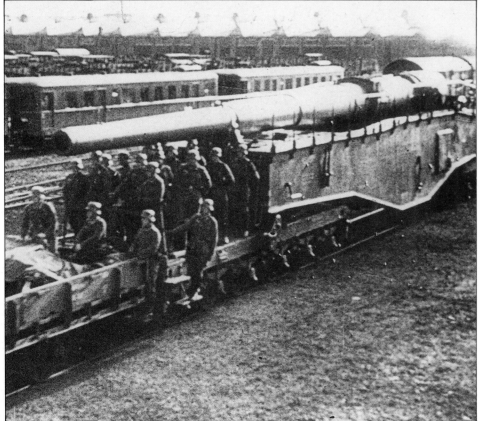

Above: A German 24cm 'Kurze Bruno' railway gun, attached to its portable turntable by means of hydraulic buffers that allowed the carriage a small amount of recoil on the turntable.

Left: German 28cm K5 railway gun being shunted through a railway yard en route to battle, some time in 1942.

Below: One of the two 15in guns installed near to Dover in 1942 to close the English Channel to German shipping.

**Another heavy gun extemporization of 1918, an American 12in coast defence gun on a French railway mount.**

Consider, for example the German 21cm K12(E) long-range railway gun. Design on this weapon began well before the war, and it is generally agreed that it was for no better reason than that the Army wanted to go one better than the Paris Gun that had been operated by the German Navy. Theoretical studies were carried out in the 1920s and early 1930s, and the same technique as had been used on the Paris Gun, that of a few deep grooves and ribbed shells, was re-adopted. Small barrels, of 10.5cm calibre, were built and tested, and in 1935 work began on the full-sized barrel 33.3m (109ft) long. The mounting was a steel box carried on two subframes, which were in turn mounted on double bogies with a total of 36 wheels. The gun was mounted in a cradle trunnioned to the steel box and recoiled in the cradle; in addition, the whole box unit recoiled across the subframes, controlled by a separate recoil system. The gun was so long that it had to be externally braced to prevent bending under its own weight, and in order to allow the gun to recoil without striking the ground, the box carriage was jacked up one metre from its subframe before firing.

The enormous weapon, which weighed 302 tonnes (297.3 tons) and measured 41.3m (136ft) in length, was issued to the German Army in 1939; it was well received but the business of jacking it up and down was disliked, so Krupp built a second gun with a complex hydro-pneumatic apparatus to balance the gun's weight and allow the trunnions to be closer to the breech, which permitted the gun to elevate and fire without the need for jacking. The gun went into service in the summer of 1940. No more were made, and

these two guns are only known to have been employed for a short period in late 1940 to early 1941 when they occasionally shelled Kent from positions on the French coastline. One shell, easily identified by its peculiar ribbed construction, landed at Rainham, close to Chatham, a point 88km (55 miles) from the nearest part of the French coast. A representative from Krupp, questioned postwar, is said to have remarked that although the gun was a waste of time as a practical weapon, as a technical exercise it was worth every penny of its enormous cost.

An even greater waste of time and money was the 80cm 'Gustav Gerät' railway gun, conceived by Krupp in answer to a question raised by Hitler when visiting the firm in 1936. Hitler had asked what size of weapon would be needed to defeat the French fortifications in the Maginot Line. In fact, the German Army had asked the same question in the previous year and Krupp had postulated guns of 70, 80 and 100cm calibre. With Hitler asking the same question, Krupp assumed that a nod was as good as a wink, and he forthwith set his design staff to develop the 80cm idea. Drawings were prepared and laid before the Army Weapons Office in 1937. These were approved, and manufacture began.

The production of this monster was so difficult that Krupp's schedules went wrong; instead of having the gun ready in 1940, as promised, it was actually 1942 before it could be assembled and fired at test. It was then sent to Sevastopol to join the siege there, moved to Leningrad – where the siege finished before it could be got ready – and does not appear to have been used again. Parts were discovered

by US troops in Bavaria in 1945 but the complete weapon was never found.

Gustav was so huge that it had to be moved in pieces and erected on site; a double railway line acted as the base, the bogie units were shunted into place, a gantry crane erected, and the superstructure of the gun then assembled. The two halves of the barrel were connected, the breech ring assembled, and the barrel placed in its cradle and the whole assembled to the mounting. The whole performance occupied several hundred men for three weeks. Once assembled it stood 11.6m (38ft) high, was 7.01m (23ft) wide and 43m (141ft) long and weighed 1,350 tonnes (1329 tons). The gun barrel was 32.4m (106ft) long and it fired a 4,800kg (10,582lb) shell to a range of 47km (29 miles) or a 7,100kg (15,653lb) shell to a range of 38km (24 miles). It was priceless as a propaganda instrument, but as a practical weapon of war it was pure nonsense.

While monster weapons of this sort were being put together at great expense and to no very good purpose, German engineers could be found developing guns that made excellent sense. The German airborne invasion of Crete in 1941, for example, revealed the development of recoilless guns for the first time and led to similar developments taking place in both Britain and the USA. The principle of the recoilless gun is very basic, and is best seen in the first practical example, which was developed in 1916 by Commander Davis of the US Navy; he simply put two guns back-to-back and fired them simultaneously so that each cancelled out the recoil from the other. In practical form the gun consisted of two barrels with a single chamber between them; the

49

The British 3.7in experimental recoilless gun of 1945, an early version of the weapon that took the role of the heavy anti-tank guns.

**Top: American troops demonstrating their two anti-tank guns in 1942; the 37mm M3A1 and the 3in M5.**

**Bottom: American troops of the 94th Infantry Division firing a 3in anti-tank gun across the River Rhine near Krefeld, April 1945.**

charge was fired in the chamber and it ejected a projectile from one barrel and an equal weight of buckshot and grease from the other. Since the weights and velocities of the two were equal, the gun did not recoil.

The Germans carried this to its ultimate, in that they 'ejected', from the rear, a stream of gas at high velocity, derived from the explo-

sion of an over-sized charge. The momentum of the gas was the same as the momentum of the shell, and the gun did not recoil. The drawback was that since much of the propellant gas escaped out of the back of the gun, the shell did not have either the velocity or the range that it would have had from a conventional gun of the same size.

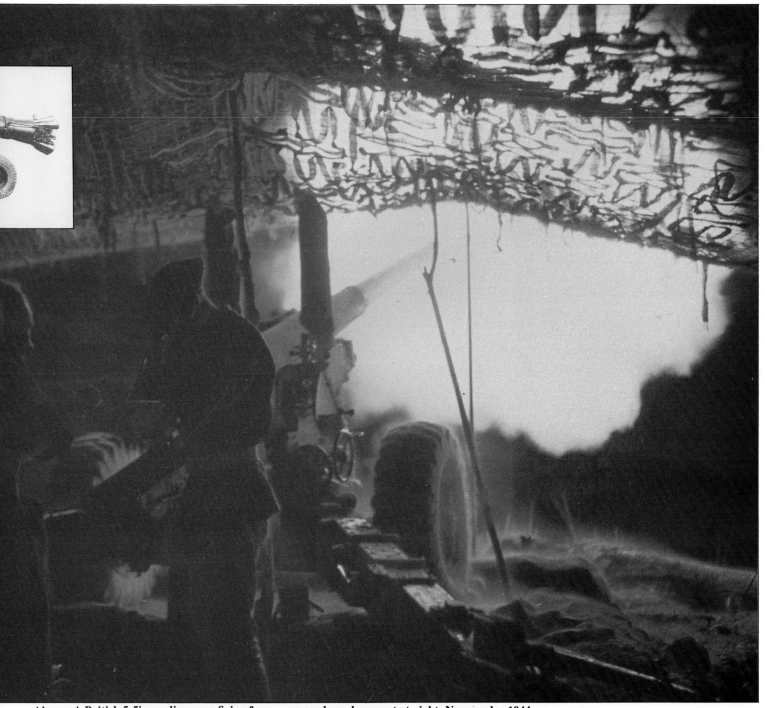

**Above: A British 5.5in medium gun firing from a covered emplacement at night; Normandy, 1944.**

The German recoilless gun was developed as a method of delivering power from a lightweight weapon that could be air-lifted. Lightness came from the fact that no recoil meant no recoil mechanism and a much less robust carriage. The reduction in weight found applications elsewhere in due course, and particularly in the field of anti-tank weapons.

The war began in 1939 with most of the belligerents fielding anti-tank weapons of 37-40mm calibre, which were just about adequate to deal with the tanks of the day. But tanks rapidly became heavier, adopted thicker armour, and began carrying guns that were more powerful than the anti-tank guns ranged against them, so in self-defence the anti-tank

gun had to grow. The German Army had a 50mm gun before the war, Britain followed with a 57mm gun, then both went to 75mm and the Germans jumped to 88mm. The Americans began with 37mm, went to 57mm and 76mm, then to 90mm and were toying with a 105mm gun when the war ended. Britain developed a 94mm gun, while Germany again

A German assault gun, the Sturmgeschütz III, developed to provide mobile supporting fire for infantry, but which subsequently found good employment as a tank-destroyer.

**American troops in the Philippine Islands using an anti-tank gun against Japanese field fortifications, May 1945.**

topped all offers by developing two 128mm models. But the defect in all this was that the calibre went up because it was becoming vital to fire heavier and heavier projectiles at higher and higher velocities, needing more powerful cartridges. As a result the guns got heavier, until they were in the eight to ten ton class. And at this weight they lost their usefulness on the battlefield, because an anti-tank gun needed to be capable of being manoeuvred rapidly by manpower, and hidden quickly for ambush purposes, which was impossible with a weapon (to take the German 128mm as an example) weighing 10,160kg (22,399lb) and with a barrel 7.02m (23ft) long. It was in response to this giantism that the recoilless gun was first explored as an anti-tank weapon. This idea was made possible by advances in ammunition design, which meant that instead of relying on a hard projectile flung at high velocity, the anti-tank shell now carried a specialized charge of explosive and would perform at low velocities.

**Left: American 'Howitzer Motor Carriage M8', a light tank mounting a 75mm howitzer in its turret, in Germany 1945.**

The other major innovation of the Second World War was the self-propelled gun, an idea that had first been tried by the French in 1918, when the Schneider and St Chamond companies developed tracked motorized mountings for heavy artillery, but had languished in the postwar years. Britain developed the 'Birch Gun' in the late 1920s, an 18pr field piece mounted in a form of turret on a Vickers tank hull. It was a reasonable design, but tribal disputes about who should operate it, the Royal Artillery or the Royal Tank Corps, led to its abandonment by 1930.

There are two approaches to the self-propelled gun, which are exemplified by the different designs produced by Germany on the one hand and Britain and the USA on the other. In Germany it was seen as a cheap replacement for the tank to accompany infantry in the assault, giving direct heavy support, and consequently it became known as the 'assault gun'. Britain and the USA, on the other hand, regarded self-propelled mountings simply as a method of getting their standard artillery pieces into position more rapidly, particularly across country, and thus allowing them to give more intimate support to fast-moving

armoured divisions. To enable the allies to retain their field weapons, the standard 25pr gun and 105mm howitzer were fitted into open-top self-propelled chassis derived from existing tanks. The Americans went further and produced mountings for their supporting 155mm guns and howitzers and even went as far as preparing designs for an 8in gun and a 240mm howitzer, though the war ended before they could be put into production. Both Allies also looked at the assault gun but declined it in favour of the 'tank destroyer' (a tank chassis carrying a powerful frontally mounted anti-tank gun), which had also originated in Germany. The Allies preferred to give their guns more scope and developed a turreted design that resembled a tank but carried a far more powerful gun – a 76mm then a 90mm in American service, a 76mm 17-pounder in British use.

Tank destroyers, though, were found to have their drawbacks. The idea was sound; send a vehicle with a heavy gun into an ambush position, kill the approaching tank or tanks, then motor rapidly to another ambush and repeat the performance. But the German pattern of frontal gun, with limited movement to its flanks, invited infantry to sneak around behind and blast the vehicle with mines. The Allied pattern, with rotating turret, tended to encourage the vehicle commander in the belief that he had some sort of tank, and far too many tried to go hunting tanks with the result that they became the victims.

Artillery reached its zenith during the Second World War, in quantity, quality and employment. The wartime development of the nuclear bomb and the guided missile led to a rethinking of the utility of the gun, and to some ill-considered verdicts. Coast defence artillery, for example, was abandoned by the Americans in 1948 and by the British in 1956, the reason being that modern missiles made nonsense of fixed shore defences. Yet in 1987 many countries are still employing coast defence artillery and the Scandinavian countries are actually designing new turret installations and mobile coast defence guns. The long-range gun and the railway gun have both vanished, supplanted by the guided missile or free-flight rocket, which can deliver a nuclear warhead to far greater range than even the Paris Gun and with much better accuracy.

But the immediate support of the infantry soldier still relies upon the gun and the howitzer, though the proliferation of types of years gone by is no longer to be found. Almost every country outside the Soviet bloc uses 105mm and 155mm guns and howitzers, and ammunition from one country can be fired in guns of another, so far has standardization been achieved. In the Soviet bloc there is similar agreement on calibres, though these differ from the Western standards, using 122mm and 152mm as its principal weapons. But the course of artillery from 1945 has shown some interesting fashions.

**Firing the 105mm Light Gun during a demonstration at the Royal School of Artillery.**

**Far right: A Puma helicopter lifting an American 105mm M101 howitzer during a NATO exercise in Portugal, 1981.**

**Above: British troops manning American 175mm self-propelled guns during a ceremonial parade before Her Majesty the Queen at Dortmund, 1984.**

**Right: The Swiss Oerlikon twin-barrelled 35mm towed air defence gun.**

**Left inset: A British 5.5in gun in use with the Baluchistan Artillery in Oman, 1972.**

**Left: Airborne gunners of 7th Regiment, Royal Horse Artillery, firing the 105mm Light Gun on Salisbury Plain.**

Most nations had sufficient wartime artillery left over to be able to live on their fat for several years after the war; what seemed to obsess the designers during that period was improvement in tanks and the beginnings of the guided missile. Ten years after the war had ended the only new piece of artillery equipment that the British Army had received was a thermometer for measuring the temperature of the propellant charges. But the changing aspects of war demanded some new designs, and the first move was to develop self-propelled weapons that would not only protect the crews from splinters and bullets but also from nuclear radiation and fall-out. Then came a

period when air mobility was all, and in the USA came a series of SP guns that was little more than stripped-down chassis with cannon on top, capable of being lifted in huge aircraft.

Next came the helicopter era, and the SP gun was forgotten for a while as efforts were made to develop powerful field pieces that could be lifted by ordinary troop-carrying helicopters. Thus the British 105mm Light Gun and the US 105mm M102 are both pared of any excess weight but can manage to range 15-17km (9-11 miles) quite accurately.

**A Belgian Para-Commando howitzer battery on NATO exercises.**

The medium and heavy anti-aircraft gun vanished in the 1950s, replaced by guided missiles, though the Soviet bloc still retains a large number which must be assessed as being of doubtful value. But the arrival of the helicopter has roused interest in the development of light anti-aircraft guns, and guns of 23-35mm are now being produced all over the world, their utility enhanced by the adoption of radars and sophisticated computers that can detect, track and predict targets with an accuracy never dreamed of during the last war.

**Bottom left: Belgian Para-Commandos firing their 105mm howitzer M101.**

**Bottom right: Baluchistan artillery firing a British 25pr gun in Oman.**

**Above: The Italian 105mm M56 pack howitzer being fired in the anti-tank role by British troops.**

Left: The Indian 75mm pack howitzer.

Below: The tri-national FH-70 155mm howitzer deployed on a training exercise.

The anti-tank gun 'proper' vanished in the early 1950s, victim of its own weight, and was replaced almost universally by the recoilless gun. Now this is being phased out and replaced by wire-guided missiles. But the late 1980s are seeing a revival of the anti-tank gun idea, lightweight guns of 75-90mm calibre taking advantage of light alloys and advances in ammunition technology to project armour-piercing projectiles at high velocity but without the weight penalty that appeared inseparable from it in 1945.

Artillery may not be so sophisticated as the guided missile or the jet-propelled bomber, but it still has one valuable property; it is the

**Top: The South African 155mm self-propelled howitzer G6 firing.**

**Above: The French 105mm TR gun towed behind a light truck.**

**Right: The tri-national FH-70 155mm howitzer firing.**

**Above:** An Italian design of twin 40mm air defence gun, turret-mounted and capable of being towed into position.

**Right:** The British twin 6pr 'intermediate' anti-aircraft gun, a development that never achieved success.

only guaranteed all-weather, 24-hour battle-field fire support system that cannot be inter-fered with by electronic means. Once the shell leaves the gun, it is on a ballistic trajectory, and nothing short of divine intervention can prevent it landing where it was aimed. In these days of electronic countermeasures and coun-ter-countermeasures, this simple assurance is of great comfort to military commanders. And this alone is enough to ensure that artillery will continue to serve for another few centuries as well as it has for the past six and a half.

British gunners resting between fire missions with their 105mm light gun.

**Top:**
The South African 155mm self-propelled howitzer G6, an unusual design that uses wheels instead of tracks.

The American 155mm self-propelled gun T97, developed in 1952, was one of a number of designs intended to protect the crew from nuclear effect as well as the more common battle hazards.

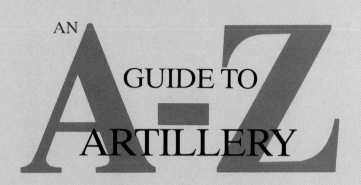

AN A-Z GUIDE TO ARTILLERY

# A

**A TUBE** In the case of a built-up gun (ie, one constructed by shrinking a number of tubes or hoops on top of each other), the 'A' Tube is the innermost tube of all, and is the actual barrel of the gun with the rifling and chamber formed on the interior. The succeeding tubes are then called 'B', 'C' and 'D' and so on, the last letter indicating the outer tube. In some cases, particularly in larger guns, it is thought advisable to make the 'A' Tube removable without having to expand and remove the hoops above it so that when the rifling is worn a fresh tube can be inserted. Cheaper and quicker than building a complete new barrel, the process entails the smooth inner surface of the "A" Tube being lined with a tapered 'Inner "A" Tube' of slightly larger dimensions (.005 inch) which is forced into position and thereby places the Inner 'A' Tube into a state of compression similar to that which would have been obtained by shrinking the 'A' Tube over the inner tube. An Inner 'A' Tube contributes to the strength of the gun; a 'liner' (qv) does not. See 'Built-up gun', 'Hoops'.

**ABBOTT** A British self-propelled 105mm field artillery gun, introduced in 1964 and currently in service with the British and Indian armies. It consists of a tracked chassis carrying a turret in which the 105mm gun is installed, giving 360° of traverse and 70° of elevation. Powered by a Rolls-Royce multi-fuel engine the vehicle has a road speed of almost 50km/hr (31mph). The gun fires a 16kg (35.3lb) shell to a maximum range of 17km (10.6 miles), using an eight-zone propelling charge system; this gives the gun a variety of charge options for much of its fighting range, permitting it to be used as a gun or as a howitzer. In addition to the standard high explosive shell, the range of ammunition includes screening smoke, signalling (coloured) smoke, squash-head (anti-tank) and illuminating (star) shells. Six of the 40 rounds carried on the mounting are normally anti-tank rounds. The maximum rate of fire for short periods is 12 rounds per minute.

**ACHILLES** A British self-propelled anti-tank gun (or tank destroyer), used 1944-50. The American M10 had its original issue 76mm gun removed and replaced by the more powerful British 17pr. The vehicle was based on the chassis of the M4A2 Sherman tank and had the gun mounted in an open-topped turret capable of 360° traverse. It was a highly effective weapon.

**The British 105mm Abbot self-propelled gun.**

**ADJUST** To 'adjust fire' means to order corrections to the aim of guns based upon observation of the fall of shot around the target. The initial round of an engagement is fired from map data or from an estimation by the observer. Once this lands the data is then 'adjusted' until the shells are hitting the target. See 'Observed fire'.

**ADOLF** During the Second World War the German Army held two different artillery equipments known by the name of 'Adolf'. It is probable that the dual naming stemmed from the use of the same gun in two different applications. The gun in question was the 40.6cm Krupp SK C/34 weapon developed in 1937 for 'H' Class battleships which, in the event, were never built. Seven guns had been built by the time the ship design was cancelled in 1939, and they were then taken for use in coast defences. Four were sent to Norway in 1942, one being lost en route, and installed close to Harstad to protect the approaches to Narvik. The other three were first built as railway guns (below) but were dismantled and installed as 'Batterie Lindemann' at Sangatte, on the Pas de Calais from where they were able to bombard south-eastern England and cover the English Channel. Two of the Narvik guns were in a twin turret; the rest were in single-gun turrets, generally protected by massive concrete casemates. The gun fired a 610kg (1,345lb) long-range shell propelled by 363kg (800lb) of powder, or a 1,030kg (2,271lb) shell propelled by a 301kg (664lb) charge. The maximum range was 56km (35 miles) with the long-range shell. The installations were retained by the Norwegian Army after the war as active coast defence weapons and were kept until the mid 1960s.

In 1938 three barrels were taken for installation on railway mounts, called the 40.6cm Kanone (Eisenbahn). The intention was to fit the guns to the mountings already under development for the 38cm 'Siegfried' railway guns (qv), but the extreme size of the weapon led to problems in fitting everything within the loading gauge and production was delayed. Three were eventually completed in 1940 and two were sent to the German occupation army in Poland, where they were installed on the Hel peninsula to protect Danzig. It was soon realized that they were superfluous in this role and, as noted above, they and the third gun, retained in Germany, were sent to France where the guns were removed and installed in turrets near Sangatte. In railway gun form, Adolf weighed 323 tonnes (318 tons), measured 31.3 metres (103ft) between buffers, and had a maximum range of 56km (35 miles) using the same ammunition as the coast artillery gun.

**Below: The two-gun Adolf turret under construction in the Krupp works.**

**AIMING CIRCLE** An American term for an optical angle-measuring instrument used for simple survey tasks and for alignment of guns. See 'Director', which is the British and French term.

**An American aiming circle, an optical instrument for orientating guns or performing simple survey tasks.**

**AIMING POINT** When engaged in indirect fire (qv) the gun is not within sight of its target and thus cannot aim at it. In order to point the gun correctly, an arbitrary 'aiming point' is selected such as a church tower, a prominent house or some other immobile and unmistakable feature and the angle between this and the target is measured. If the gun's panoramic sight (qv) is now set at this angle and the sight pointed at the aiming point by moving the gun, the muzzle of the gun will point at the target.

**AIMING RIFLE** This was originally a gun of small calibre (usually 1 inch [25mm]) which could be inserted into the barrel of a larger-calibre gun so that its bore was parallel with the bore of the parent weapon, and fired for short-range target practice. Aiming rifles could also be attached to the outside of the gun barrel for the same purpose.

In present-day practice, aiming rifles are .50in calibre self-loading rifles or machine guns attached to the barrel of anti-tank recoilless guns and provided with special ammunition which duplicates the trajectory of the parent weapon's shell. The aiming rifle is fired by the gunner at varying ranges until a hit on the target is obtained; the parent gun is then fired and a hit will usually be obtained, since the aiming rifle shots will have established the correct elevation and corrected for wind or other transient conditions. Now generally called 'spotting rifles'.

**AIR BLAST** A compressed air apparatus that is connected to the breech ring of heavy coast artillery and fortress guns. Supplied from a tank of air, the blast is operated immediately after firing and before opening the breech to eject any smoke or fumes from the gun barrel. It also acts as a forced draught and inflames any smouldering residue from the cartridge bag which may be in the chamber, ejecting the remains from the muzzle. Without air blast, opening the breech would permit quantities of smoke and toxic fumes to enter the gun house, turret or casemate, to the detriment of the gun detachment, and any smouldering residue might be fanned into flame by the draught as the breech was opened. Present-day practice is to use a fume extractor (qv) or bore evacuator (qv) rather than air blast.

**AIRBURST** The practice of bursting shells in the air above or close to the target by use of time or proximity fuzes in the shell allows shell fragments to be driven downwards and thus penetrate vertical cover – eg, a wall or trench – which would otherwise protect the enemy from the fragments of a shell bursting on the ground. Airburst shells can also be used to check the accuracy of the gun or to determine a correction to the gun data to compensate for meteorological conditions. A point of burst is calculated and the gun is fired in accordance with the calculated data; the actual point of burst is observed by surveying instruments and its position exactly calculated. This is then compared with the theoretical position and corrections to the gun data derived from the comparison.

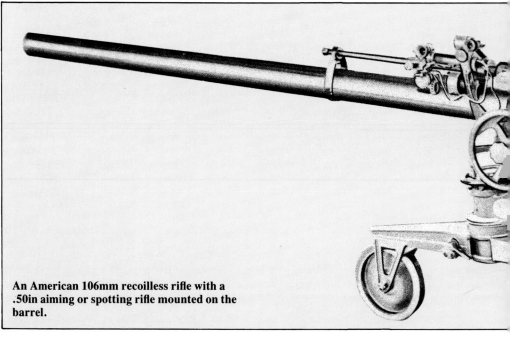

An American 106mm recoilless rifle with a .50in aiming or spotting rifle mounted on the barrel.

The British Alecto SP equipment, mounting a 95mm 20-calibre howitzer, was intended for airborne use but did not enter service.

**ALECTO** British self-propelled howitzer developed in 1943-4 but never adopted for service. It used a 95mm howitzer barrel mounted in an open-topped hull built on the chassis of the Harry Hopkins light tank, with a view to producing a light weapon that could be air-transported. There were difficulties with the production of the 95mm recoil system, and the project was abandoned early in 1945.

**AMERICAN ARTILLERY** In spite of a well-developed heavy engineering industry, there has been relatively little manufacture of artillery outside the US government arsenals. During the latter part of the 19th century the Driggs-Seabury Company and the Bethlehem Steel Company manufactured small numbers of field and coast defence guns for export sale and for supply to the US Army and Navy but, in general, the government purchases were due to shortfall in the production from the arsenals and were not on a regular contractual basis.

The design and development agency for US artillery has always been the Ordnance Depart-

**A battery of 90mm M2A2 guns on a firing range in North Germany, 1956.**

ment, and with a succession of autocratic Chiefs of Ordnance in the latter part of the 19th century the US Artillery had very little say in what equipment they used. Design and manufacture of gun barrels was largely carried out at Watervliet (qv) Arsenal, in New York state. The design and development of carriages and final assembly of complete guns was usually done by Rock Island Arsenal, Illinois. The manufacture of some coast artillery mountings, of the larger and more complex kind, was often contracted out to major engineering companies, though the designs originated with the Ordnance Department.

As a result of this policy, and also because of the reluctance of Congress to authorize funds, the actual introduction of guns into service lagged several years behind their design and approval dates. Although a major rearmament of coast defences, for example, was authorized in 1886 and many designs were approved in 1888, it was not until the latter years of the 1890s that the guns actually began to appear in service. It is for this reason that Driggs-Seabury, Bethlehem, and British Armstrong and Vickers guns appeared in US service in small numbers in the 1892-1905 period.

The US Ordnance Department was working on a

new 3 in field gun, the M1916, when the First World War broke out. This gun proved almost impossible to mass-produce when the US entered the war in 1917, and the Army was forced to obtain supplies of French 75mm M1897 guns, which it later manufactured under licence and continued to use into the 1940s. At the same time they bought quantities of British 18pr field guns, which were being manufactured on British contracts in the USA, and redesigned them to accept the French 75mm ammunition; a similar modification was done on the 3in M1916 gun. Similarly, there was very little heavy field artillery in existence in the USA, and thus

## US ARTILLERY EQUIPMENT

| Equipment | Date | Calibre (mm) | Barrel length (cals) | Breech mech. | Elevation max (deg) | Traverse (deg) | Weight in action (kg) | Shell weight (kg) | Muzzle velocity (m/sec) | Range, max (metres) |
|---|---|---|---|---|---|---|---|---|---|---|
| **Anti-tank artillery** | | | | | | | | | | |
| 37mm Gun M3 | 1938 | 37 | 54 | VSB | 15 | 60 | 414 | 0.7 | 883 | 11750 |
| 57mm Gun M1 | 1941 | 57 | 50 | VSB/SA | 15 | 90 | 1274 | 2.9 | 853 | 11585 |
| 3in M5 | 1941 | 76.2 | 50 | HSB | 30 | 45 | 2212 | 5.8 | 853 | 14000 |
| 90mm T8 | 1944 | 90 | 52 | HSB/SA | 20 | 60 | 3084 | 11 | 853 | 19600 |
| | | | | | | | | | | |
| **Anti-aircraft artillery** | | | | | | | | | | |
| 37mm M1 | 1938 | 37 | 54 | VSB/SA | 90 | 360 | 2778 | 0.6 | 792 | 5670 ceiling |
| 75mm M51 'Skysweeper' | 1953 | 75 | 40 | VSB/A | 85 | 360 | 9357 | 6.8 | 853 | 9150 ceiling |
| 3in M1917A2 | 1917 | 76.2 | 55 | VSB/SA | 85 | 360 | 6800 | 5.8 | 853 | 8960 ceiling |
| 3in M3 | 1928 | 76.2 | 50 | VSB/SA | 80 | 360 | 5535 | 5.8 | 853 | 9450 ceiling |
| 3in M4 | 1931 | 76.2 | 55 | VSB/SA | 85 | 360 | 6800 | 5.8 | 853 | 8850 ceiling |
| 90mm M1 | 1940 | 90 | 50 | VSB/SA | 80 | 360 | 8035 | 10.6 | 822 | 9750 ceiling |
| 105mm AA M3 | 1927 | 105 | 60 | VSB/SA | 80 | 360 | 15212 | 14.9 | 853 | 12800 ceiling |
| 120mm Gun M1 | 1943 | 120 | 60 | VSB/SA | 80 | 360 | 22135 | 22.7 | 945 | 17500 ceiling |

**Right: American troops of the 2nd Infantry with a 37mm anti-tank gun in Iceland, 1942.**

**Left: Night firing with a 37mm AA gun early in World War Two.**

French 155mm howitzers and guns and British 6in guns and 8in and 9.2in howitzers were bought in some numbers.

In spite of severely restricted finance, much work was carried out designing new guns during the 1920s to replace these 1917 adoptions, though few prototypes were made. When war threatened in 1940, however, these designs were ready and were able to go into production with minimum delay. All the World War One designs had been replaced by 1943, though a number of them stayed in use as training weapons, and many new designs had been added. In the tables that follow, the notation 'M1917' or some other full date, indicates a World War One era weapon, while the notation 'M1' or a numerical model designation shows that the design dates from the 1940s.

The development of self-propelled guns was a particular feature of the US Second World War artillery programme, as might be expected from a country with a strong automotive industry. The first to appear in service was the 155mm M12, using the M1917 gun barrel on a tracked chassis, but other designs came thick and fast after it. Many achieved no more than a prototype and are not mentioned here, but before the war ended serviceable equipments mounting the 8in gun and 240mm howitzer had been developed.

| US Artillery Equipment (Cont) Equipment | Date | Calibre (mm) | Barrel Length (cals) | Breech mech. | Elevation max (deg) | Traverse (deg) | Weight in action (kg) | Shell weight (kg) | Muzzle velocity (m/sec) | Range max (metres) |
|---|---|---|---|---|---|---|---|---|---|---|
| **Mountain artillery** | | | | | | | | | | |
| 2.95in Gun M1903 | 1903 | 75 | 12 | SQF | 27 | Nil | 375 | 5.6 | 280 | 4480 |
| 3in Howitzer M1911 | 1912 | 76.2 | | | 40 | 6 | 500 | 6.8 | 275 | 5200 |
| 75mm Pack How M1 | 1935 | 75 | 19 | HSB | 45 | 5 | 576 | 6.8 | 380 | 8400 |
| 75mm Pack How M116 | 1955 | 75 | 19 | HSB | 45 | 6 | 653 | 6.8 | 384 | 8280 |
| | | | | | | | | | | |
| **Field artillery** | | | | | | | | | | |
| 3in M1902 | 1902 | 76.2 | 29 | SQF | 15 | 8 | 970 | 6.8 | 518 | 6860 |
| 3in M1916 | 1918 | 76.2 | 28 | VSB/SA | 53 | 45 | 1380 | 6.2 | 600 | 8800 |
| 75mm M1916 | 1917 | 75 | 28 | VSB/SA | 53 | 45 | 1456 | 6.1 | 579 | 11420 |
| 75mm M1917 | 1918 | 75 | 30 | SQF | 16 | 8 | 1300 | 6.8 | 530 | 8000 |
| 75mm M1897A2 on Carr M2A2 | 1934 | 75 | 40 | NS | 46 | 85 | 1575 | 6.8 | 665 | 11685 |
| 75mm M1897A4 | | 75 | 36 | NS | 19 | 6 | 1365 | 6.2 | 575 | 6335 |
| 75mm Gun M2A2 | 1940 | 75 | 40 | NS | 46 | 85 | 1560 | 6.7 | 662 | 13695 |
| 75mm How M1 on Carr M3A1 | | 75 | 19 | HSB | 49 | 55 | 855 | 6.7 | 380 | 8800 |
| 105mm How M2A1 | 1939 | 105 | 22 | HSB | 66 | 46 | 2030 | 14.9 | 472 | 11200 |
| 105mm How M3A1 | 1942 | 105 | 18 | HSB | 49 | 55 | 1134 | 14.9 | 310 | 7585 |
| 105mm How M102 | 1965 | 105 | 30 | VSB | 76 | 360 | 1496 | 14.9 | 494 | 11500 |
| 105mm SP How M108 | 1965 | 105 | 30 | HSB | 74 | 360 | 22450 | 14.9 | 472 | 12000 |
| 4.7in How M1908 | 1908 | 120 | | | 40 | | 2175 | 27.2 | 275 | 6250 |

*US Artillery Equipment (Cont)*

| Equipment | Date | Calibre (mm) | Barrel Length (cals) | Breech mech. | Elevation max (deg) | Traverse (deg) | Weight in action (kg) | Shell weight (kg) | Muzzle velocity (m/sec) | Range max (metres) |
|---|---|---|---|---|---|---|---|---|---|---|
| **Heavy artillery** | | | | | | | | | | |
| 4.7in Gun M1906 | 1906 | 119 | 29 | SQF | 15 | 8 | 3250 | 27.2 | 520 | 8750 |
| 4.7in Gun M1917 | 1917 | 119 | 29 | SQF | 40 | 8 | 3660 | 20.4 | 710 | 15000 |
| 4.5in Gun M1* | 1941 | 114 | 43 | IS | 65 | 53 | 5475 | 24.9 | 690 | 18750 |
| 5in Gun M1898 | 1900 | 127 | 27 | IS | 31 | | 1650 | 20.3 | 550 | 9000 |
| 5in Gun M1918 | 1918 | 127 | 34 | IS | 21 | 8 | 4475 | 27.2 | 650 | 11000 |
| 155mm Gun M1918M1 | 1918 | 155 | 38 | IS | 35 | 60 | 11300 | 43 | 735 | 18380 |
| 155mm Gun M1 | 1938 | 155 | 45 | IS | 63 | 60 | 12700 | 43.5 | 853 | 23500 |
| 155mm How M1917 | 1917 | 155 | 15 | IS | 43 | 6 | 3750 | 42.8 | 450 | 11500 |
| 155mm How M1 | 1941 | 155 | 23 | IS | 63 | 50 | 5765 | 43.1 | 565 | 15000 |
| 155mm How M114A1 | 1960 | 155 | 23 | IS | 63 | 48 | 5760 | 43.1 | 564 | 14600 |
| 155mm SP How M109 | 1965 | 155 | 29 | IS | 75 | 360 | 23796 | 43.1 | 561 | 14600 |
| 155mm How M198 | 1979 | 155 | 39 | IS | 72 | 45 | 7163 | 43.1 | | 22000 |
| 6in How M1908 | 1908 | 152 | 13 | SQF | 40 | 6 | 3350 | 54.4 | 275 | 6150 |
| 175mm SP Gun M107 | 1963 | 175 | 60 | IS | 65 | 30 | 28170 | 66.8 | 923 | 32700 |
| 7in How M1890 | 1893 | 178 | 12.7 | IS | 40 | | | 47.6 | 330 | 5500 |
| 8in How M1 | 1940 | 203 | 25 | IS | 65 | 60 | 14380 | 90.7 | 595 | 16925 |
| 8in How M115 | 1965 | 203 | 25 | IS | 65 | 60 | 13472 | 90.7 | 594 | 16800 |
| 8in Gun M1 | 1943 | 203 | 50 | IS | 50 | 30 | 31435 | 108.8 | 865 | 32330 |
| 240mm How M1918 | 1920 | 240 | 18.7 | IS | 60 | 20 | 18734 | 156.5 | 518 | 15000 |
| 240mm How M1 | 1943 | 240 | 35 | IS | 65 | 45 | 29350 | 163.3 | 700 | 23065 |
| 280mm Gun M65 | 1952 | 280 | 45 | IS | 55 | 360 | 42638 | 272 | 762 | 28700 |

**Superheavy: the 240mm Howitzer M1.**

The 8in Gun M1 used the same mounting as the 240mm howitzer and was the 'long range sniper' of the pair.

At the end of the war many designs that were about to go into production were dropped, on the grounds either that the development, while satisfactory for wartime purposes, was not sufficient for peacetime, or that the particular requirement that had given rise to the design had now evaporated; eg, the super-heavy self-propelled guns, which had been specifically developed for the assault on Japan.

Many of the wartime designs were slightly overhauled in postwar years and given new model numbers – the M1 105mm howitzer became the M101, for example – and new designs were developed in response to changing tactical and strategic ideas. The first major shift came with the development of turreted SP guns intended to protect their crews from nuclear flash and blast; then the pendulum swung in the other direction and the 1960s saw many designs of gun and howitzer in which air-portability was the pre-eminent requirement. The present US artillery armoury is heavily biased to the self-propelled gun, largely because of the Army's bias towards armoured divisions, but in recent years the resurgence of the light infantry concept and the Rapid Deployment Force has seen more development work on lightweight artillery. In 1985 the British L119 105mm Light Gun was adopted (as the M119) for employment by the US Marines and for light division of the US Army.

**One of the 16in coast defence guns guarding Pearl Harbor. Unfortunately, the Japanese struck from the air – not the sea.**

| *US Artillery Equipment (Cont)* Equipment | Date | Calibre (mm) | Barrel Length (cals) | Breech mech. | Elevation max (deg) | Traverse (deg) | Weight in action (kg) | Shell weight (kg) | Muzzle velocity (m/sec) | Range max (metres) |
|---|---|---|---|---|---|---|---|---|---|---|
| **Coast artillery** | | | | | | | | | | |
| 2.24in RF Gun M1900 | 1900 | 57 | 50 | VSB | 12 | 360 | | 2.7 | 731 | 5700 |
| 3in Gun M1902 | 1902 | 76.2 | 50 | SQF | 15 | 360 | | 6.8 | 792 | 7955 |
| 3in Gun M1903 | 1903 | 76.2 | 55 | SQF | 16 | 360 | 4213 | 6.8 | 853 | 10350 |
| 4in Driggs-Schroeder | 1900 | 101 | 40 | SQF | 15 | 360 | | 14.9 | 700 | 8100 |
| 4.72in Armstrong | 1895 | 120 | 49 | SQF | 20 | 360 | | 20.4 | 792 | 10250 |
| 5in Gun M1900 | 1900 | 127 | 50 | SQF | 15 | 360 | | 26.75 | 792 | 10780 |
| 6in RF Armstrong | 1896 | 152 | 40 | SQF | 16 | 360 | | 48 | 655 | 9325 |
| 6in M1897 | 1900 | 152 | 45 | IS | 15 | 170 | | 40.8 | 792 | 10800 |
| 6in Gun M1900 | 1903 | 152 | 50 | IS | 20 | 360 | 20670 | 40.8 | 838 | 15100 |
| 6in Gun M1903 | 1905 | 152 | 50 | IS | 15 | 170 | | 40.8 | 838 | 15500 |
| 6in Gun M1903A2 | 1940 | 152 | 50 | IS | 47 | 360 | 72125 | 47.6 | 853 | 24825 |
| 6in Gun M1905 | 1908 | 152 | 50 | IS | 47 | 360 | 72125 | 47.6 | 853 | 25150 |
| 6in Gun M1908 | 1910 | 152 | 45 | IS | 15 | 120 | 19165 | 40.8 | 792 | 13525 |
| 6in Gun M1 | 1942 | 152 | 50 | IS | 47 | 360 | | 47.6 | 853 | 24825 |
| 8in Gun M1888 | 1896 | 203 | 32 | IS | 18 | 360 | 40155 | 146.5 | 670 | 14900 |
| 8in Gun Mk VI Mod 3A2 | 1937 | 203 | 45 | IS | 42 | 360 | 71670 | 108.8 | 838 | 32580 |
| 10in Gun M1888M1 | 1891 | 254 | 34 | IS | 15 | 320 | 65590 | 231.4 | 685 | 14900 |
| 10in Gun M1900 | 1902 | 254 | 40 | IS | 12 | 170 | 180530 | 280 | 685 | 14900 |
| 12in Mortar M1890 | 1892 | 305 | 10 | IS | 70 | 360 | 71215 | 475 | 517 | 10990 |
| 12in Mortar M1908 | 1910 | 305 | 10 | IS | 65 | 360 | 65410 | 475 | 305 | 8400 |
| 12in Mortar M1912 | 1913 | 305 | 15 | IS | 65 | 360 | 74980 | 317 | 550 | 17665 |
| 12in Gun M1888 | 1895 | 305 | 34 | IS | 15 | 360 | 103875 | 442 | 681 | 16825 |
| 12in Gun M1895 | 1898 | 305 | 35 | IS | 35 | 360 | 184480 | 442 | 688 | 27525 |
| 12in Gun M1900 | 1902 | 305 | 40 | IS | 10 | 360 | 305450 | 485 | 685 | 15850 |
| 14in Gun M1907M1 | 1908 | 355 | 34 | IS | 20 | 360 | 288490 | 752 | 716 | 20850 |
| 14in M1909 in Turret ** | 1913 | 355 | 40 | IS | 15 | 360 | 1050537 | 547 | 722 | 20830 |
| 14in M1910 | 1913 | 355 | 40 | IS | 20 | 170 | 309355 | 752 | 716 | 20850 |
| 16in Gun M1895 | 1917 | 406 | 35 | IS | 20 | 170 | 577882 | 1088 | 685 | 25025 |
| 16in Gun M1919MII | 1923 | 406 | 50 | IS | 30 | 360 | 566092 | 1061 | 838 | 38040 |
| 16in Gun M1919 | 1928 | 406 | 50 | IS | 65 | 360 | 492156 | 1016 | 822 | 45720 |
| 16in Howitzer M1920 | 1927 | 406 | 25 | IS | 65 | 360 | 408240 | 952 | 594 | 22400 |

Two 12in Seacoast Mortars M1890 in a typical mortar pit at Fort Monroe, Virginia, 1917.

| US Artillery Equipment (Cont) Equipment | Date | Calibre (mm) | Barrel Length (cals) | Breech mech. | Elevation max (deg) | Traverse (deg) | Weight in action (kg) | Shell weight (kg) | Muzzle velocity (m/sec) | Range max (metres) |
|---|---|---|---|---|---|---|---|---|---|---|
| **Railway artillery** | | | | | | | | | | |
| 8in Gun M1888M1 | 1918 | 203 | 32 | IS | 42 | 360 | 71550 | 117.9 | 792 | 21850 |
| 8in Gun Mk VI Mod 3A2 | 1940 | 203 | 45 | IS | 45 | 360 | 104328 | 117.9 | 838 | 32275 |
| 10in Gun M1888M1 | 1918 | 254 | 34 | IS | 54 | Nil | 175630 | 231.4 | 731 | 21950 |
| 12in Mortar M1890 | 1918 | 305 | 10 | IS | 65 | 360 | 80200 | 317 | 457 | 13980 |
| 12in Gun M1895† | 1919 | 305 | 35 | IS | 38 | 5 | 146060 | 317 | 693 | 27450 |
| 12in Gun M1895†† | 1918 | 305 | 35 | IS | 40 | Nil | 249480 | 317 | 975 | 41150 |
| 14in M1919 | 1919 | 355 | 40 | IS | 30 | 10 | 260820 | 635 | 716 | 38850 |
| 14in M1920M1 | 1920 | 355 | 50 | IS | 50 | 7 | 331130 | 545 | 914 | 44100 |
| | | | | | | | | | | |
| **Recoilless artillery** | | | | | | | | | | |
| 57mm RCL Rifle M18 | 1945 | 57 | 27 | SQF | 65 | 360 | 20 | 1.2 | 365 | 4500 |
| 75mm RCL Rifle M20 | 1945 | 75 | 27 | SQF | 65 | 360 | 76 | 6.5 | 300 | 6675 |
| 90mm RCL Rifle M67 | 1955 | 90 | 15 | SQF | 65 | 360 | 35 | 3.1 | 215 | 750 |
| 106mm RCL Rifle M40A1 | 1965 | 105 | 32 | SQF | 65 | 360 | 219 | 7.7 | 498 | 6875 |
| 120mm RCL M28 Davy Crockett | 1958 | 120 | 12 | ML | 45 | 360 | 47 | | 140 | 2000 |
| 155mm RCL M29 Davy Crockett | 1958 | 155 | 16 | ML | 45 | 360 | 141 | | 200 | 4000 |

The 14in railroad Gun M1920, of which four were built in the 1920s. Two were held in Panama and two near San Francisco, for coast defence purposes.

**Left:** An M109 155mm howitzer of the US 25th Infantry Division, firing near Cu Chi, Vietnam, in 1968.

**Lower left:** An 8in howitzer M110 moving into position during exercises in Germany, 1968.

**Right:** A 175mm M107 SP gun at Landing Zone Elliot, Vietnam, 1969.

**Far right:** 'Sergeant York', a twin 40mm self-propelled equipment which proved unsuccessful and was withdrawn in 1985.

## US SELF-PROPELLED ARTILLERY

| Equipment | Date | Calibre (mm) | Chassis | Elevation max (deg) | Traverse on mount (deg) | Weight in action (kg) | Range, max (metres) | Ammunition carried (rds) |
|---|---|---|---|---|---|---|---|---|
| M3 GMC | 1941 | 75 | Half track | 29 | 40 | 9075 | 1000 (A/Tk) | 59 |
| M6 GMC | 1942 | 37 | 4×4 truck | 15 | 360 | 3335 | 1000 (A/Tk) | 80 |
| M7 HMC | 1941 | 105 | M3 tank | 30 | 15 | 3650 | 10600 | 36 |
| M8 HMC | 1942 | 75 | M5 tank | 40 | 360 | 15700 | 8790 | 46 |
| M10 GMC | 1942 | 76 | M4 tank | 19 | 360 | 29935 | 14600 | 54 |
| M12 GMC | 1941 | 155 | M3 tank | 30 | 28 | 26750 | 20100 | 10 |
| M18 A/Tk GMC | 1944 | 76 | Special | 20 | 360 | 17035 | 3000 (A/Tk) | 45 |
| M19A1 AA GMC | 1944 | 40 twin | M24 tank | 85 | 360 | 17450 | 3350 (AA) | 360 |
| M36 A/Tk GMC | 1944 | 90 | M10 SP | 20 | 360 | 28120 | 15600 | 47 |
| M37 HMC | 1945 | 105 | M24 tank | 43 | 45 | 20870 | 11165 | 90 |
| M40 GMC | 1945 | 155 | M4 tank | 55 | 36 | 37660 | 23520 | 20 |
| M41 HMC | 1945 | 155 | M24 tank | 45 | 37 | 18595 | 14960 | 22 |
| M42A1 AA GMC | 1953 | 40 twin | M41 tank | 85 | 360 | 22450 | 5000 (AA) | 480 |
| M43 HMC | 1945 | 203 | M4 tank | 52 | 34 | 37660 | 16930 | 16 |
| M44 HMC | 1953 | 155 | M41 tank | 65 | 60 | 28350 | 14600 | 24 |
| M52 HMC | 1951 | 105 | M41 tank | 65 | 120 | 24050 | 11270 | 102 |
| M53 GMC | 1951 | 155 | Special | 65 | 60 | 43550 | 23225 | 20 |
| M55 HMC | 1951 | 203 | Special | 65 | 60 | 44465 | 16800 | 10 |
| M56 A/Tk GMC | 1957 | 90 | Special | 15 | 60 | 7035 | 2500 (A/Tk) | 29 |
| M107 GMC | 1961 | 175 | Special | 65 | 60 | 28170 | 32700 | 2 |
| M108 HMC | 1961 | 105 | Special | 74 | 360 | 22450 | 11500 | 87 |
| M109 HMC | 1963 | 155 | Special | 75 | 360 | 23800 | 14600 | 18 |
| M109A2 HMC | 1978 | 155 | Special | 75 | 360 | 24950 | 18100 | 36 |
| M110 HMC | 1961 | 203 | Special | 65 | 60 | 26500 | 16800 | 2 |
| M110A2 HMC | 1978 | 203 | Special | 65 | 60 | 28350 | 21300 | 2 |
| T48 | 1942 | 57 | Half track | 15 | 55 | 8620 | 1000 (A/Tk) | 99 |
| T92 | 1945 | 240 | Special | 50 | 22 | 58530 | 23100 | nil |

**Notes:** AA — Anti-aircraft; A/Tk — Anti-tank; GMC — Gun Motor Carriage; HMC — Howitzer Motor Carriage; * British 4.5in BL gun adapted carriage ** Weight of turret assembly with two guns † Rolling mount †† Sliding mount RF Rapid Fire (American equivalent to 'Quick firing')

The US 4.7in heavy AA gun, which was developed during World War Two but was never used.

**ANGLE OF DEPARTURE** The angle between the horizontal plane and the axis of the gun bore when fired; it incorporates quadrant elevation and jump.

**ANGLE OF DESCENT** The angle between the horizontal plane and the line of arrival of the shell as it strikes the ground.

**ANGLE OF INCIDENCE** The angle between the surface of the target and the line of arrival of a projectile; eg, an anti-tank shot striking a vertical armour plate. In British practice the angle is considered 0° (or 'normal') when the shot strikes the target at a perfect right-angle, and the angle then increases as the line of arrival becomes more oblique. In Continental practice the shot striking at right-angles is considered to have an angle of incidence of 90°, the angle being measured from the face of the target. Thus a shot striking at 30° by British standards is striking at 60° by Continental standards, and this must be borne in mind when comparing the performance figures of different weapons. The current NATO practice is to adhere to the Continental standard.

**ANGLE OF PROJECTION** The angle between the line of sight and the axis of the bore when the gun is fired; consists of Tangent Elevation plus jump.

**ANGLE OF SIGHT** The angle between the horizontal plane and a sight line connecting the gun to the target; may be an angle of elevation or of depression. Known in American service as 'Site'.

**ANTI-AIRCRAFT ARTILLERY** Artillery specifically developed to attack aircraft in flight. It is generally characterized by the ability to traverse unrestrictedly, to elevate to at least 80°, have a high velocity (in order to get the shell to the target in the shortest possible time, so that the target movement will be minimal) and fire as rapidly as possible.

The first anti-aircraft (AA) guns were developed in 1908-9 in Germany, by Krupp and Rheinmetall, and were known as 'balloon guns'. They were 65-75mm weapons, sometimes mounted on motor vehicles since it was thought that the gun would be able to pursue a slow-moving balloon or airship. Most of the combatant nations had developed some form of AA gun by the outbreak of war in 1914, but these guns existed only in small numbers. The rapid development of aviation during that war was echoed by an equally rapid development of AA guns in calibres up to about 120mm. Fire control formed the most difficult area of the AA problem. It was first dealt with by fitting complex sights to the guns, and later by having a 'central post sight' in the middle of a group of guns, where the firing data would be worked out and then ordered to the guns. Ammunition was another area in which much development was needed, particularly in the provision of time fuzes which allowed the point of burst of the shell to be predicted with some accuracy.

The basic AA problem is that the target is moving at high speed in three dimensions. The shell is set on its course from the instant the gun fires and cannot be directed, so that any movement of the target subsequent to the firing has to be estimated – or perhaps 'guessed' would be a better word. As a

The British 3.7in static-mounted AA gun with full electric power control.

result, there arose the convention of assuming that the target would continue flying at the same speed and height and on the same course; without this assumption, there would have been little hope of any rational solution to the problem. Having determined the position of the target, it was necessary to determine its course, speed and height and calculate gun data which would place a shell in the same space as the target. The next step was to calculate how far the target would have moved during the time of flight of the shell, and then re-calculate firing data to deliver the shell at the future position of the target.

In the early days this had to be done by charts and tables, but in the 1920s the first 'predictors' – mechanical computers – were devised to solve this problem rapidly and continuously. Electronic computers, faster and capable of greater accuracy, arrived in the 1940s.

The AA gun was developed during the 1920s and a division appeared between light AA and heavy AA; light AA was of small calibre (20-40mm), had a very high rate of fire (100-120 rounds per minute) and generally dispensed with predictors in favour of computing sights (qv). This class of weapon invariably used impact-fuzed shells which also contained self-destroying devices that blew them up in the air should they miss, so that the live shell did not fall to earth. Heavy AA meant weapons from 75mm calibre upwards, allied with predictors and time-fuzed ammunition. During the Second World War further classification appeared; guns of 75-100mm became medium AA, and anything above that was heavy. A new class, the intermediate AA gun, was devised, attempting to cover that part of the sky which was too high for the light AA gun but too low for the medium and heavy guns. Britain and Germany both tried to develop 55-57mm guns to meet this requirement, but without success.

The principal Second World War advances in AA gunnery came with the adoption of powered loading and fuze-setting, which speeded up the rate of fire; the adoption of mechanical time fuzes which improved the practical ceiling (qv); and, eventually, the development of proximity fuzes that required no setting (so speeding up loading and firing), detected the target and burst the shell if within lethal distance.

Development in the immediate postwar years, confronted with jet aircraft that flew at high speed and great height, was principally concerned with increasing muzzle velocity to reduce the time of flight, increasing the lethality of shells to give greater lethal areas, and increasing the rate of fire. However, just as these problems appeared to be within distance of solution, the guided missile completed its development, and by 1960 the medium and heavy AA gun had vanished from the West. Numbers are still retained by Warsaw Pact armies, but their worth is questionable.

The light or intermediate gun, however, still has a role to play, even though light missiles are becoming more common. Guns of 20-30mm calibre are in use with most Continental armies and the 40mm Bofors gun is still widely used. These weapons are primarily deployed for the protection of front-line troops against ground attack aircraft or for the protection of columns on the move. At the time of writing, particular attention is being paid to the helicopter as a ground attack aircraft and, as a result, a number of light AA gun designs specifically intended to combat the helicopter are under review.

**One of the heaviest and most powerful anti-aircraft guns ever built was the German 128mm FlaK 40.**

**ANTI-TANK ARTILLERY** These were weapons designed specifically to attack tanks and other armoured vehicles. They were generally characterized by high velocity, low silhouette, wide arcs of traverse, fixed ammunition and a high rate of fire. They were usually on split-trail carriages to give the widest possible arc of fire, though some designs used cruciform platforms which allowed a full circle of traverse.

The first 'pure' anti-tank gun – that is, a gun designed to attack tanks, not merely a gun designed for some other purpose and pressed into service against tanks – appears to have been a small 37mm weapon developed by the Germans in 1918, though few were made and fewer used. During the 1920-39 period most countries developed and adopted weapons in the 20-37mm class, as these allowed high velocities to be developed without excessively large rounds of ammunition, and were small enough to be manhandled and concealed by three or four men.

The Second World War brought improvements in tanks, particularly in their armour protection, so that this class of gun was soon outmoded and heavier weapons began to appear. The German Army led the way with a 50mm gun in 1940 and the British followed with a 57mm in 1941. Gradually the calibres crept up as the increasing thickness of armour demanded heavier projectiles and more powerful charges to penetrate it. By 1945 the British had a 76mm 17pr in service, the Germans an 88mm in service and the Americans were using a 90mm gun on a self-propelled mounting. The British had a 94mm under development, the Germans had a 128mm starting production, and the Americans had developed a 105mm weapon. These latter three showed the defect in anti-tank weapon development, in that they were all far too heavy to be practical weapons on the battlefield.

The attack on armour is, properly, an ammuni- tiion subject, but it must be briefly mentioned here in order to explain gun development. The basic projectile was a pointed steel shot, but it was soon found that there was a limit to the velocity at which such a shot could be delivered. Above about 2,800ft/sec (853m/sec) the steel would shatter on

striking armour, instead of piercing. Projectiles using tungsten cores were adopted as a result, and these demanded powerful cartridges in order to generate the necessary striking velocity. For this reason, the guns became too big. But in the early 1940s the shaped charge – a method of defeating armour by explosive force – was developed, fol- lowed, in Britain, by the squash-head shell, another explosive method of attack. These methods of attack did not rely on high velocity, so that it became possible to use lighter guns and, in particular, recoilless guns. Thus in the 1950s the 'pure' anti-tank gun vanished from the scene, to be replaced by recoilless guns firing explosive types of armour-defeating ammunition. A number of con- ventional anti-tank guns are still retained in the Warsaw Pact, but their efficiency against modern main battle tanks must be considered doubtful.

**Below: The zenith of conventional anti-tank guns was this German 128mm design of 1944.**

**Inset: American troops firing a 57mm anti-tank gun in the Philippines, 1945.**

Anzio Annie, the German 28cm railway gun, firing into the American beach-head.

**ANZIO ANNIE** American soldiers' name for the German 28cm railway gun (28cm K5[E]) that bombarded the beaches at Anzio in 1944. It was later captured at Civitavecchia and taken to the USA where it is now on show at the Artillery Museum of Aberdeen Proving Ground. The German troops called it 'Leopold'.

**ARCHER** This British self-propelled anti-tank gun consisted of the chassis of a redundant 'Valentine' tank from which the turret had been removed and replaced by an armoured superstructure into which a 17pr gun was fitted. For the sake of simplicity and compactness, the gun was mounted to fire over the rear of the vehicle, which meant that it had to be reversed into its firing position. Once in position the driver had to get out of his seat quickly to avoid being struck by the recoiling breech. Archer was a useful equipment, although somewhat primitive, and performed well in north-west Europe in 1944-5. A small number was retained after the war, but all were obsolete by the early 1950s.

**Above: The British Archer 17pr anti-tank self-propelled gun in action in Germany, 1945.**

**ARMSTRONG** William George Armstrong (1810-1900) was a Newcastle solicitor who took an interest in scientific and engineering matters, developed hydraulic machinery, and in 1847 established a factory at Elswick, near Newcastle-upon-Tyne. During the Crimean war he was appalled by reports of the difficulties experienced by the British forces at the Battle of Inkerman in moving an 18pr gun into place in muddy conditions. He was struck by the immense weight of the solid cast-iron smoothbore gun – some three tons – in comparison to the weight of projectile it fired. Armstrong designed a new type of gun, relying upon careful assessment of the internal pressure and design of the gun so as to provide the necessary strength without excess weight. His gun used a wrought-iron rifled barrel which was strengthened by having 'hoops' or hollow sleeves shrunk around it. It was breech-loaded, using a vent-piece of steel which dropped into a slot and was retained by a large screw. The projectile was coated with lead, which bit into the rifling and developed spin, so stabilizing the shell .

The British Army adopted the Armstrong Gun in 1858 after long trials. Armstrong assigned his patents to the Government and was appointed Superintendent of the Royal Gun Factory. Approximately 5,000 Armstrong guns of various calibres were made for use by the army and navy. The advent of armoured ships, demanding guns that could deliver more power than the Armstrong (which had a relatively weak breech) led to the Armstrong breech-loader being augmented by rifled muzzle-loaders in the 1860s, but the breech-loaders remained in service for many years. Armstrong's factory became the Elswick Ordnance Company and continued making guns for British and foreign service; it later became Armstrong-Whitworth, then Vickers-Armstrong, and the site of the Elswick factory is today within the Vickers factory where the Abbott self-propelled gun was built.

**ARTICULATION** A term relating to the suspension of a gun carriage and used almost exclusively in split-trail (qv) carriages. A split-trail means that there are four points of contact with the ground – two wheels and the two trail leg ends – and finding a piece of flat ground to get all four points level is impractical. Three-point contact will adapt to almost any ground, and articulation is therefore concerned with supporting the front end of the trail structure in such a way as to allow the wheels to take up whatever level they choose and give the trail three points of suspension. The simplest way is to form the trail front end into a horizontal pivot and insert this into the centre of the axle, so allowing flexing between axle and trail. There are several other, more complicated, ways of articulating a trail.

**Left: An experimental Vickers-Armstrong field gun of the 1920s, showing how articulation allows the trail ends to rest at different levels.**

**ASBURY BREECH** A breech mechanism for bag-charge guns that has certain specific features: 1, a Welin screw; 2, a rotating cam; 3, an operating lever that is vertical and on the right-hand side. The result is a fast-acting mechanism that can be opened or closed with one movement of the lever. On pulling the lever down, the breech screw is revolved to unlock; the cam converts the rotational movement into a turning movement and the momentum of the lever rotates the block.

**ASSAULT GUN** A form of self-propelled gun pioneered by the German Army in the Second World War. It consists of a tracked chassis with a superstructure that carries a gun of greater calibre than would be found on a tank of comparable size. The gun is mounted so that it fires forward and with a limited degree of traverse to either side. The object of the assault gun is, as the name implies, to accompany infantry during their assault on an objective and provide direct supporting fire against obstacles, strongpoints, machine-gun posts and similar targets that might otherwise hold up the infantry's advance. The German and Soviet armies embraced the concept of the assault gun with great enthusiasm. There is no doubt that the assault gun has considerable advantages in certain types of warfare, but its disadvantage is that it has a limited arc of fire, so it is relatively easy to outflank the vehicle and destroy it from the rear.

**ATOMIC ANNIE** A nickname for the American 280mm Gun, Heavy, Motorized M65, derived from the fact that this was the first artillery piece to fire an atomic projectile. Development of the gun began in 1944 with a plan for a 240mm gun and a 280mm howitzer that would be suspended between two tractors. The whole idea was dropped at the end of the war but was revived in the late 1940s when the prospect of an atomic shell was in view. The 280mm gun barrel was carried in a box-like carriage which was suspended at either end between two specially-built truck tractors, a system which gave the weapon remarkable manoeuvrability. Introduced into service in the mid-1950s, it remained until the late 1960s but was withdrawn due to the impossibility of concealing it from aerial reconnaissance and because new battlefield missiles made it reduntant. The gun weighed about 74 tons (75.2 tons) and had a maximum range of 18 miles (28.8 kilometres).

**Left: American troops in Germany loading the 280mm Atomic Cannon, which was the first artillery weapon capable of firing a nuclear shell.**

**Right: The 305mm 'Schlanke Emma' siege howitzer, here seen in action on the Italian front during World War One.**

**AUSTRIAN ARTILLERY** The Austro-Hungarian Empire had three sources of supply for artillery: the Skoda Company of Pilsen, the Artillerie Zeugfabrik (Arsenal) in Vienna, and the Böhler Company. Of these three Skoda was by far the largest supplier. The Artillerie Zeugfabrik had been the principal maker prior to the breech-loading age, but Skoda's much more modern manufacturing facilities brought it to the forefront when guns became more complicated; the Zeugfabrik did manufacture some mountain guns and produced some designs for Skoda to make. Böhler had some good designs prior to 1914 but only under the pressure of war did the Austro-Hungarian government find the funds necessary to purchase guns from them.

Skoda, of course, supplied guns to many other countries before 1914, among them China, Turkey and Britain.

## AUSTRIAN ARTILLERY EQUIPMENT

| Equipment | Date | Calibre (mm) | Barrel length (cals) | Breech mech. | Elevation max (deg) | Traverse (deg) | Weight in action (kg) | Shell weight (kg) | Muzzle velocity (m/sec) | Range, max (metres) |
|---|---|---|---|---|---|---|---|---|---|---|
| **Anti-tank artillery** | | | | | | | | | | |
| 47mm M35 Böhler | 1935 | 47 | 40 | HSB | 60 | 50 | 290 | 1.5 | 650 | 9000 |
| Noricum N-105 | 1983 | 105 | 56 | HSB/SA | | | 3600 | 3.7 | 1485 | 6400 |
| | | | | | | | | | | |
| **Mountain artillery** | | | | | | | | | | |
| 7cm Mountain M'08 | 1908 | 72.5 | 13.8 | SQF | 34 | 8 | 402 | 4.8 | 310 | 5200 |
| 7cm Mountain M'09 | 1909 | 72.5 | 13.8 | SQF | 35 | 8 | 456 | 4.8 | 310 | 5300 |
| 7.5cm Mountain M'13 | 1914 | 75 | 16 | HSB | 36 | 5 | 491 | 5.3 | 300 | 5600 |
| 7.5cm Mountain M'15 | 1915 | 75 | 15 | HSB | 50 | 7 | 620 | 6.5 | 350 | 7000 |
| 7.5cm Mountain How M'15 | 1915 | 75 | | HSB | | | | 6.5 | | 7800 |
| 10cm Mountain How M'08 | 1908 | 104 | 15 | HSB | 42 | 5 | 1233 | 14.7 | 300 | 6000 |
| 10cm Mountain How M'10 | 1910 | 104 | 15 | HSB | 70 | 6 | 1210 | 14.7 | 300 | 6000 |
| 10cm Mountain How M'16 | 1916 | 100 | 19 | HSB | 70 | 5 | 1235 | 16 | 340 | 7750 |
| 15cm Mountain How M'18 | 1918 | 150 | 13 | | 70 | 7 | 2800 | 42 | 340 | 8000 |

| *Austrian Artillery Equipment (Cont)* Equipment | Date | Calibre (mm) | Barrel Length (cals) | Breech mech. | Elevation max (deg) | Traverse (deg) | Weight in action (kg) | Shell weight (kg) | Muzzle velocity (m/sec) | Range max (metres) |
|---|---|---|---|---|---|---|---|---|---|---|
| **Field artillery** | | | | | | | | | | |
| 37mm Infantry gun M'15 | 1915 | 37 | 10 | VSB | 45 | 20 | 118 | 0.5 | 185 | 2400 |
| 7.5cm FK M'05 | 1905 | 75 | 30 | HSB | 18 | 6 | 950 | 6.7 | 500 | 5500 |
| 7.5cm FK M'12 | 1912 | 75 | 29 | HSB | 16 | 7 | 940 | 6.5 | 500 | 6000 |
| 7.6cm FK M'18 | 1918 | 76.5 | 30 | HSB | 45 | 8 | 1330 | 8.0 | 500 | 10500 |
| 8cm FK M'05 | 1905 | 76.5 | 30 | HSB | 18 | 8 | 1020 | 6.7 | 500 | 7300 |
| 8cm FK M'17 | 1917 | 76.5 | 30 | HSB | 45 | 8 | 1386 | 6.7 | 500 | 9900 |
| 10cm Fd How M'99 | 1899 | 104 | 13 | NS | 42.5 | 6 | 992 | 14.3 | 290 | 5575 |
| 10cm Fd How M'14 | 1914 | 100 | 19 | HSB | 48 | 5 | 1420 | 11.5 | 420 | 8000 |
| 10.4cm Fd How M'14 | 1914 | 104 | 18 | HSB | 45 | 7 | 1250 | 14.7 | 320 | 7800 |
| 10.5cm Fd How M'15/T | 1915 | 105 | 18 | HSB | 70 | 6 | 1397 | 16 | 350 | 7750 |
| 10cm Gun M'15 | 1915 | 104 | 35 | HSB | 30 | 6 | 3020 | 17.4 | 680 | 13000 |
| **Heavy artillery** | | | | | | | | | | |
| 12cm Siege Gun M'80 | 1880 | 120 | 27 | HSB | 30 | | 3640 | 17.5 | 515 | 8000 |
| 15cm Siege Gun M'80 | 1880 | 149 | 24 | HSB | 28 | | 5510 | 33.1 | 480 | 8500 |
| 15cm Siege How M'80 | 1880 | 149 | 8 | | 65 | | 2000 | | 205 | 3500 |
| 15cm Fd How M'94 | 1895 | 149 | 13 | | 65 | | 2470 | 38.3 | 300 | 6200 |
| 15cm Fd How M'14 | 1914 | 149 | 14 | HSB | 70 | 8 | 2765 | 42 | 350 | 8000 |
| 15cm Fd How M'15 | 1915 | 149 | 20 | HSB | 65 | 8 | 5560 | 42 | 510 | 11500 |
| 15cm Fd How M'38 | 1938 | 149 | 24 | HSB | 65 | 45 | 5630 | 41 | 580 | 14050 |
| 15cm Gun M'15 | 1915 | 152.4 | 34 | HSB | 32 | 6 | 12200 | 56.5 | 700 | 19000 |
| 15cm How M'15 | 1915 | 149 | 14 | | 70 | 8 | 2500 | 38.3 | 340 | 8100 |
| Noricum GH-45 | 1979 | 155 | 45 | IS/SA | 72 | 70 | 10070 | 46 | 897 | 30300 |
| Noricum M114/39 | 1982 | 155 | 39 | IS/SA | 63 | 49 | 6500 | 46 | 785 | 21000 |
| 18cm Siege Gun M'80 | 1880 | 180 | 12 | HSB | 35 | | 4350 | 58 | 260 | 5100 |
| 25cm 'Gretel' How | 1898 | 240 | 9 | IS | 65 | | 7040 | 132.5 | | 6500 |
| 28cm How | 1914 | 280 | 12 | HSB | 65 | 10 | 15765 | 338 | 340 | 11000 |
| 30.5cm 'Emma' How | 1913 | 305 | 14 | IS | 70 | 120 | 20000 | 380 | 340 | 12000 |
| 30.5cm How M'16 | 1916 | 305 | 12 | | 75 | 360 | 23000 | 380 | 380 | 12300 |
| 35cm Railway M'16 | 1916 | 350 | 45 | HSB | 50 | 360 | 205000 | 700 | | 30000 |
| 38cm Railway 'Lulo' | 1917 | 380 | 40 | | | | | 850 | 795 | 38000 |
| 38cm How M'16 | 1916 | 380 | 17 | | 75 | 360 | 83000 | 1000 | 459 | 15000 |
| 42cm How L/15 | 1917 | 420 | 15 | | 70 | 360 | 106000 | 1000 | 415 | 14600 |
| **Coast artillery** | | | | | | | | | | |
| 21cm Coast Mortar M'80 | 1880 | 210 | 11.4 | HSB | 45 | 360 | | 95 | 295 | 7200 |
| 30.5cm Coast/Rly gun | 1915 | 305 | 40 | HSB | 23 | 180 | | 455 | 700 | 18000 |

The Skoda factory found itself in Czechoslovakia in 1919, and continued to manufacture guns for export, supplying them to Austria, Germany, Bulgaria, Hungary, Poland and Romania as well as to the Czech Army. Austria continued to use its pre-war Skoda guns, plus one or two postwar models, and relied principally on improved ammunition to bring the older weapons up to modern standards. In 1938, with the annexation of Austria into the Third Reich, the army standardized on German weapons, though Skoda mountain guns were still held and used. After 1945 the Austrian Army continued to use German equipment for some years, but then adopted American weapons.

**AUTOCANNON** A French anti-aircraft gun developed in 1914, it consisted of a 75mm field artillery gun in a special high-angle mounting, placed in the rear of a De Dion Bouton automobile. It was used extensively in France, but a small number was purchased by Britain and used in the London Air Defences in 1915-16.

Inset: A French 75mm Autocannon mounted on a De Dion chassis. Several of these were acquired for London in 1915.

The 75mm Autocannon in firing position.

**AUTO-FRETTAGE** A system of gun construction in which the gun barrel or 'A' tube is pre-stressed by filling it with liquid and then subjecting it to hydraulic pressure that is considerably greater than the internal pressure it will be expected to stand in service. This expands the metal of the barrel; the inner layers expand beyond their elastic limit and take up a permanent set, while the outer layers are expanded below their elastic limit and therefore contract when the auto-frettage pressure is removed. The inner layers are thus placed in a state of permanent compression that is analogous to shrinking on an outer hoop, so that the structure of the gun is strengthened. Auto-frettaging allows a given degree of strength to be obtained in a lighter barrel, or, for the same weight of barrel, allows the gun to be stronger. Also called 'Cold-working'.

The credit for first suggesting the idea is claimed for a French artillery officer, L. Jacob, in 1907. Another Frenchman, Malaval, a naval ordnance engineer, published a paper in 1912 which claims originality, but a little-known paper by L.B. Turner of King's College Cambridge, read to the Cambridge Philosophical Society in 1909, contains a complete mathematical treatment of the subject and sets out proposals for the auto-frettage of gun barrels in a clear manner. By 1913 the French had manufactured a 14cm gun, and during the First World War the Military Superintendent of the British Royal Gun Factory at Woolwich made contact with French experts and instructed Major A.E. MacRae RA, Assistant Superintendent of the Design Department at Woolwich, to investigate the possibilities. MacRae conducted exhaustive research for some ten years and produced a complete solution to all the problems, details of which were published in his book *Overstrain of Metals* in 1930. From that time onwards the manufacture of gun barrels by auto-frettage developed rapidly.

**AUTOSIGHT** Contraction of 'Automatic Sight'. Type of direct-fire sight that was invented in Italy c.1875 and used with coast defence artillery as an automatic rangefinder. The height of the gun above sea level was precisely known and it follows that for any given range there must be one angle of depression, as the height of the gun and the surface of the sea are two elements of a right-angled triangle. The autosight was fitted with a cam which synchronized the movement of the gun and sight so that when the gun was elevated or depressed until the crosswire of the sight telescope aligned with the waterline of a ship, the gun had the correct elevation to hit the target. Corrections were required, of course, to compensate for the rise and fall of the tide.

**An autosight for a British 4in coast defence gun. As the gun 'a' elevates, the sight movement is controlled by the curved arm riding in a curved cam. The cam is moved by a lever to compensate for the rise and fall of the tide.**

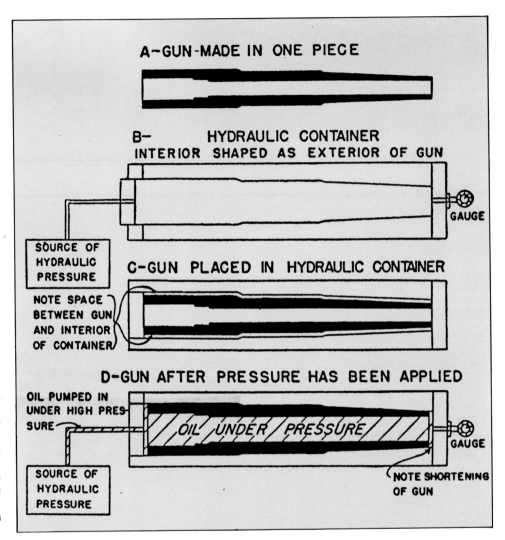

**A diagram showing the process of autofrettage: filling the gun with oil under pressure to induce a permanent set in the metal.**

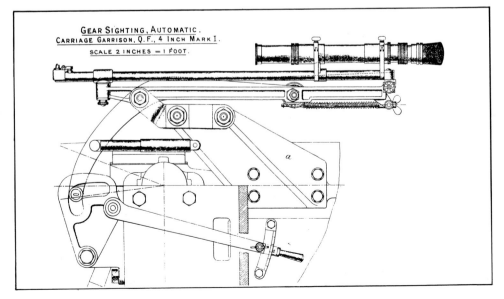

**AUXILIARY PROPULSION** The fitting of a motor unit to a gun carriage to drive the gun wheels and permit the gun to be manoeuvred under power without the need for a tractor. The motor unit can be an 'add-on' unit clamped to the trail and driving a gun-wheel via a flexible drive, or it can be engineered into the gun carriage and drive both wheels by gears or hydraulic motors. The trail ends need to be supported on a wheel or wheels which must be steerable.

An example of the former type is the Soviet 85mm Gun SD-44. Examples of the latter include the FH70, Bofors FH77 and French TR 155mm howitzers. Probably the first auxiliary propelled gun was an experimental British 17pr anti-tank gun, developed in 1944 by Nicholas Straussler of the Alvis Company. This had a widened carriage and mounted extra wheels at the trail end. The gun wheels were driven by a Ford truck engine carried on the trail, and a complete set of driving controls was mounted alongside the gun breech. The equipment could also tow its own ammunition trailer. The idea worked, but was turned down because the enlarged carriage made it extremely difficult to conceal or dig into a pit.

# B

**BALANCED PILLAR** A type of gun mounting used with coast and fortress artillery. The gun is mounted in a normal pedestal type of mounting which, in turn, is placed on top of a steel pillar or cylinder. This is counter-balanced so that it can be raised or lowered inside a circular shaft in the floor of the gun emplacement. The raising and lowering can be done mechanically, hydraulically or by hand winches. The pillar is normally lowered into the shaft to lower the gun into a position of concealment behind the parapet of the gun emplacement. When required for action, the pillar is raised to lift the gun mounting until the barrel is clear of the parapet and can engage targets in front. The pillar remains up throughout the engagement; it does not sink between shots, and in this respect the balanced pillar differs from the disappearing mount. It is not found with guns larger than 15cm/6in calibre, for reasons of weight. The balanced pillar was used by America, British colonial and other coast defences and in some European land fortresses.

**Above: A 9.2in coast defence gun in a barbette emplacement, showing how the mounting is concealed behind a parapet over which the gun fires.**

**Right: A diagram of an Armstrong balanced pillar mounting for a 4.7in coast defence gun.**

**BALLISTIC PENDULUM** A measuring device used in experimental ranges, now obsolete. Invented by Benjamin Robins (qv), it consisted of a pendulum arm suspended in a massive framework and carrying a large and solid mass at its lower end. A gun under test fired a solid projectile at the mass, and the impact caused the pendulum to swing. This swing was measured and, from considerations of the length of the pendulum arm, weight of the mass and distance moved, it was possible to calculate the striking energy of the projectile. Knowing the weight of the projectile it was possible to determine the striking velocity. Once accurate methods of measuring velocity were available, the ballistic pendulum was no longer necessary, but in the 18th and 19th centuries it was a valuable experimental tool.

**BARBETTE** A type of gun mounting used with fixed fortress artillery in which the gun mounting is concealed in an emplacement pit and only the gun muzzle and upper shield can be seen above the front parapet by the enemy. The term comes from the French word 'Barbe' (beard), and is said to derive from the emplacement parapet's resemblance to a beard beneath the gun barrel.

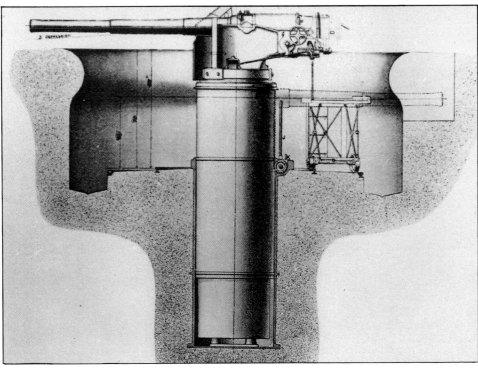

**BARRAGE** A method of artillery fire in which the shells fall in a line, so forming a barrier in front of the advancing infantry or in front of any area it is required to protect. The line of the barrage can be static (a 'standing barrage') or it can move, by increasing the elevation on the guns at regular intervals (a 'moving barrage'). If the guns are divided into different groups and fire so that one group fires one line, the other group fires a line ahead, and the first group 'leapfrogs' to fire the next line ahead and so on, the result is a 'rolling barrage'.

**BAT** An abbreviation for 'Battalion Anti-Tank gun'; adopted by various countries, but principally it applies to the 120mm recoilless gun developed in Britain in the late 1940s.

During the war a series of recoilless guns had been developed by Sir Denis Burney (qv), but at the end of the war it was felt that there was still a great deal that was not known about these weapons and the whole project went back into the research stage, the work being done at the Royal Armament Research and Development Establishment, Fort Halstead. As a result of this study, the Burney idea of a perforated cartridge case was abandoned and the German system of using a plastic seal to the cartridge case was adopted. In this system the cartridge case has a metallic base formed into a venturi, and this is then closed by a heavy plastic disc. There is a corresponding hole in the breech-

block of the gun. On firing, the plastic disc holds long enough to allow the projectile to start up the bore, after which it bursts and allows gas to flow through the venturi and the breechblock and exhaust to the rear of the gun, so balancing the recoil. The advantage of this system is that the venturi is new, and of the correct (critical) dimension for each round; with the Burney gun the venturi, being part of the gun, wore away with each shot until it failed to balance and the gun was actually 'recoiling' forward on firing.

The BAT was issued in 120mm calbire in the late

1950s. In subsequent years it was modified into the 'Wombat' and 'Mobat' versions, these being merely improved and lightened carriages, with improved sights and the addition of an aiming rifle. The projectile fired by the BAT was always a squash-head armour-defeating shell, based on Burney's 'wall-buster' design and using plastic explosive to defeat armour.

The BAT was declared obsolete in 1984, although a small number are still held by the British Army garrison in Berlin; it has been generally replaced by the Milan anti-tank guided missile.

**Left: A British 120mm BAT recoilless anti-tank gun deployed in Belize, 1975.**

**BATIGNOLLES MOUNT** A type of railway gun mounting developed by the Schneider-Creusot company of France and named after the Batignolles factory where the first mounting was built. It consists of a steel box structure that rides on the usual sort of wheeled bogies at each end and carries the gun in a cradle. To place the weapon into the firing position, the location is first selected, the track is removed and a prefabricated ground platform of steel cross- and side-girders is laid. With the platform completed, the track is relaid across the top and the gun pushed into position by a locomotive. Blocks and wedges are driven between the girders of the platform and the under-surface of the gun mounting box structure to take the weight from the wheels, and the mounting is clamped securely to the platform. The recoil of the gun is completely absorbed by the recoil system on the mounting, and the firing shock is transferred to the ground through the blocks and girder platform. Two platforms were usually carried with one gun, so that two positions on an épée (qv) track could be prepared. It was used with a number of French and American railway guns of the First World War period.

**BATTERY** 1, A specific group of guns forming an administrative and tactical unit. There are usually six or eight guns per group, though this can vary; a single large-calibre railway gun could be a battery on its own.
2, The position of the gun barrel in the mounting when ready to fire. After the gun recoils, it runs back 'into battery'.

**The American 12in M 1895 gun on a Batignolles railway mounting, showing how the wedges were applied to take the weight from the wheels.**

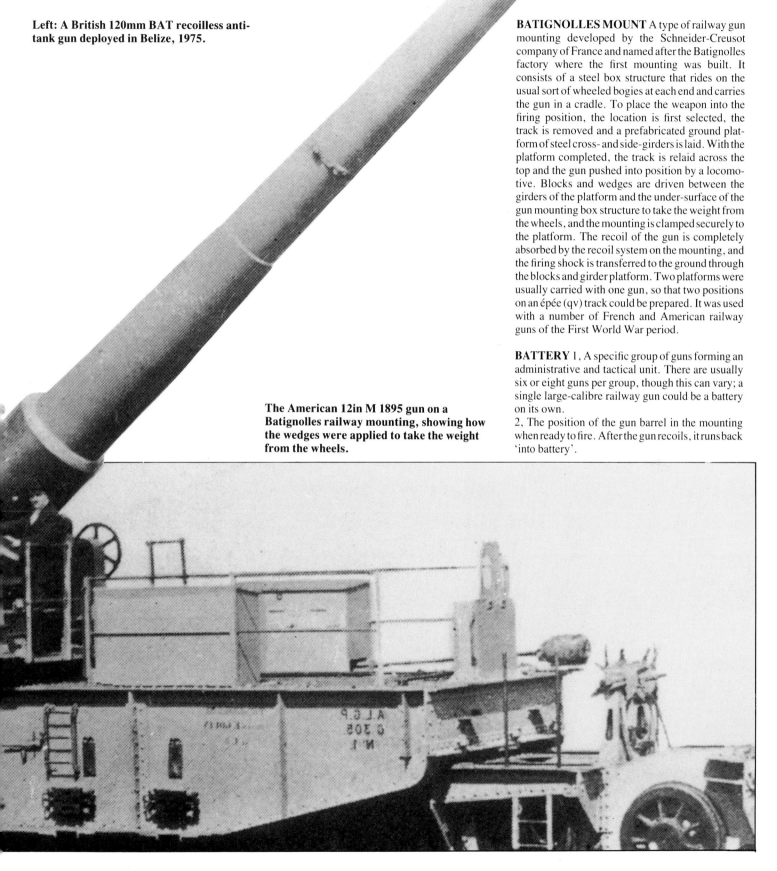

**BELGIAN ARTILLERY** There were two gun-making firms in Belgium, Cockerill and FRC, but the Belgian Army bought its guns from Germany and France. Some of these were manufactured by the two Belgian firms under license, but otherwise the private makers built for export, supplying guns to China, South America and the Balkans. The First World War saw the addition of numbers of 75mm French M1897 field guns to the Belgian stocks, but after the war the Belgians reverted to purchasing from Krupp and Rheinmetall. After 1945 the Army had virtually no stocks of any kind of weapon, and since that time American equipment has been used almost exclusively, the only other major purchase being of Italian 105mm M56 pack howitzers.

**BELL-MOUTH** A gun is said to be 'bell-mouthed' when the bore is widened for the last inch or so of the muzzle. The idea was first applied to rifled muzzle-loading guns in order to make the insertion of the shell easier when loading, but it was also found to reduce the formation of cracks around the muzzle by reducing pressure on the thinnest part of the barrel. The practice was abandoned with muzzle-loading but has been revived in recent years after it was discovered that modern high-strength steel shells have a tendency to stretch the last portion of the gun chase during their passage, thus causing cracks at the muzzle.

**BETHLEHEM SCREW** A type of breechblock developed by the Bethlehem Steel Corporation in the 1890s. It was an interrupted screw block in which the screw-threads were of 'involute' form; that is, their diameter altered on a gentle curve instead of being stepped. When locked the screw had a bearing area of some 240° and was extremely strong, but the manufacture was difficult and expensive and the design did not meet with much success.

**BETHLEHEM STEEL CORPORATION** American steel company which manufactured guns and carriages in the period 1890-1918. It manufactured field and naval guns to its own design prior to the First World War, selling them to the US Army and Navy and abroad. During the war it manufactured many British designs of guns and howitzers to meet British contracts, and later manufactured the same weapons to meet US Army requirements.

**BIG BERTHA** A nickname given to the Krupp 42cm howitzers used to bombard the fortress of Liège in August 1914, the name being derived from Baroness Bertha Krupp, daughter of Alfried Krupp the gunmaker. The 42cm was based on an earlier design of 42cm coast defence howitzer in which the German Army had expressed interest, provided it could be made portable, and which Krupp redesigned as a mobile weapon. It was transported piecemeal behind five motor tractors and assembled on site by a crane. When in action it weighed 42 tonnes (41.3 tons), and fired a shell weighing 816kg (1,719lb) to a maximum range of 9,375 metres (10,253 yards). Two howitzers were built and were used against several fortresses in Belgium and France and also on the Russian front. They were withdrawn from service late in 1917 due to their short range, which had been surpassed by Allied counter-battery weapons. Examples were not found after the war and it is assumed that they had all been scrapped. It should be noted that Big Bertha was *not* the gun that bombarded Paris in 1918.

**BIRCH GUN** A British self-propelled gun developed in 1925, it consisted of an 18pr field gun mounted in a turret on a tracked chassis derived from the then-current Vickers medium tank. In 1926 a second version was built in which the gun was mounted without a turret so that it could elevate to 85° and thus act in a secondary anti-aircraft role. In 1927 one battery of six guns was issued to the Experimental Mechanized Force for trials and appears to have been successful. In 1928 a third version was proposed, in which the gun was to be shielded and available for the field role only, the high angle requirement being dropped, but in the same year the Mechanized Force was disbanded and the Birch Gun project was abandoned. The gun took its name from General Sir Noel Birch, then Master-General of the Ordnance.

**BISHOP** The first British self-propelled gun to be employed in the Second World War. Used in 1942-3, it was a 25pr field gun placed in a large armoured box on top of a Valentine tank chassis. Due to the confines of the box the gun could not be elevated to its maximum and therefore the greatest range which could be reached was only 6,400 yards (5,852 metres). The crew was cramped, but in spite of these drawbacks the gun was reasonably successful and was used in North Africa, Sicily and in the early weeks of the Italian campaign. Its most useful function was to act as an introductory equipment to teach regiments self-propelled gun tactics.

## BELGIAN ARTILLERY EQUIPMENT

| Equipment | Date | Calibre (mm) | Barrel length (cals) | Breech mech. | Elevation max (deg) | Traverse (deg) | Weight in action (kg) | Shell weight (kg) | Muzzle velocity (m/sec) | Range, max (metres) |
|---|---|---|---|---|---|---|---|---|---|---|
| **Anti-tank artillery** | | | | | | | | | | |
| 47mm FRC | 1932 | 47 | 34 | HSB | 20 | 40 | 568 | 1.5 | 675 | 5200 |
| **Field artillery** | | | | | | | | | | |
| 7.5cm Gun M'05 | 1905 | 75 | 30 | HSB | 15 | 7 | 1030 | 6.5 | 500 | 8000 |
| 7.5cm Gun Model TR | 1912 | 75 | 30 | | 16 | 7 | 1050 | 6.5 | 500 | 8000 |
| 7.5cm Gun M'18 | 1918 | 75 | 35 | | 43 | 20 | 1450 | 7.3 | 600 | 11000 |
| 10.5cm Gun M'13 | 1913 | 105 | 28 | | 37 | 6 | 2336 | 16.3 | 548 | 12450 |
| 120mm Fd How | 1914 | 120 | | | 40 | | 1300 | 20 | 300 | |
| 12cm Gun M'32 | 1932 | 120 | 37 | HSB | 40 | 60 | 5450 | 21.9 | 770 | 18150 |
| **Heavy artillery** | | | | | | | | | | |
| 155mm Gun M'17 | 1917 | 155 | 32 | IS | 42 | 6 | 8840 | 43 | 667 | 15550 |
| 15cm How M'17 | 1917 | 150 | 17 | IS | 40 | 5 | 2286 | 41.2 | 335 | 8600 |
| 155mm How M'24 | 1924 | 155 | 30 | IS | 45 | 8 | 7840 | 43 | 665 | 17000 |

**Above: Bishop, the first British 25pr self-propelled gun of World War Two.**

**The Birch gun was the first British self-propelled gun, mounting an 18pr gun on a Vickers medium tank chassis.**

**BL** British abbreviation for 'Breech-Loading' but which is generally reserved specifically for a breech-loading gun that uses a bagged charge. A breech-loading gun using a metal cased cartridge is a 'QF' gun (qv).

**BLAKELY** English gun designer, operating about 1850-70. His system of gun construction involved using a cast-iron bore, strengthening this with a wrought iron hoop over the chamber, and then shrinking a further cast iron hoop around the whole. Alternatively, his gun could be built up from an 'A' tube, a breech-piece and an enclosing hoop which held everything together. All his designs were rifled muzzle-loaders, generally using a modification of the Woolwich system of three deep grooves allied to studs on the shell. They were given some tests in Britain but were not considered adequate for service. Blakely was able to sell a number of his guns, in calibres from 8 to 12.75 inches, to the Confederate and Union Armies during the American Civil War when they were used in coast defences and in naval service.

**BOCHE-BUSTER** A British railway gun that was operational in both World Wars. Built in 1917 by the Elswick Ordnance Company, it was a simple box structure on two pairs of bogies, carrying a 14in gun mounted in a cradle that was supported on trunnions pivoting on the side members of the mounting. The gun had been manufactured for the Japanese Navy but could not be delivered due to the war and was taken into service as the 14in Mark 3. The gun fired a 720kg (1,587lb) shell to a range of 29 kilometres (18 miles) or a 635kg (1,400lb) shell to 34.75km (21.6 miles). Located at Arras, the gun was operated by 471 Siege Battery, Royal Garrison Artillery, until the end of the war. It was declared

obsolete in 1926; the gun was scrapped, but the mounting was stored.

In 1940 the mounting was brought out and fitted with an 18in howitzer barrel which had been developed at the end of the First World War. It was stationed close to Dover and deployed for beach defences since its maximum range was not sufficient to allow it to fire across the Channel. A 'Superheavy Railway Regiment' was formed in 1943 with the intention of going to the Continent after the invasion. In the event, tactical air support was found to do all that this howitzer could have done and the idea was abandoned. The equipment was withdrawn shortly after the war ended. The

mounting was scrapped, but the howitzer is still in existence at an Experimental Range. Performance with 18in barrel: shell weight 1,135kg (2,502lb), range 20.3 miles (12.6 miles); weight of 'equipment, 254 tonnes (250 tons).

**BOFORS** Swedish gunmaking company located at the town of Bofors in Central Sweden. Gun manufacture began in 1883 with small-calibre field guns, and in the ensuing years field, medium, coastal and anti-aircraft guns have been designed and built. During the 1920s/30s the German firm Krupp held shares in Bofors and many Krupp designers were employed there during the period when gunmaking

**Left: The familiar Bofors 40mm light anti-aircraft gun at firing practice. It has been used throughout the world since 1929.**

**Above: 'HM Gun Boche-Buster' in France, 1918.**

in Germany was restricted by the Versailles Treaty. The company's most famous product, which is synonymous with their name throughout the world, is the 40mm automatic anti-aircraft gun which was developed in 1929 and, in improved form, is still manufactured. This, the 40mm L/60, was mounted on a light four-wheeled carriage and fired at a rate of 120 rounds per minute. In later years it was adapted to naval mountings and fitted to self-propelled guns by various licensees, given power-operation and remotely-controlled from radars. In the 1950s it was improved into the L/70 model, using more powerful ammunition and with the rate of fire doubled.

**BOFORS SCREW** A type of breech screw designed and patented by Bofors in the early 20th century and still employed by them in some artillery weapons. It is an interrupted screw but with a curved conical-shape block instead of having the threads cut on a regular cylinder. The shape permits the breech block to be swung open as soon as it has been unlocked, without requiring it to be axially withdrawn for a short distance before beginning the opening swing.

**BOHLER** Gebrüder Böhler AG was an Austrian engineering company that went into gunmaking c.1890 (as Böhler, Kapfenberg AG) and continued to manufacture artillery until 1945. The company still exists but is no longer concerned with armaments. Although its original products were small-calibre field guns, the company appears to have specialized in anti-tank guns in its later years and produced a 47mm weapon that was used by the Austrian, German and Italian armies in some numbers during the Second World War.

**BORE** 'The bore is the interior of the gun, extending from the muzzle face to the rear of the chamber, and includes both the chamber and the rifled portion.' (from *Textbook of Service Ordnance*, 1923)

**BORE EVACUATOR** Alternative name for 'Fume extractor' (qv) and generally used in American terminology.

**BORESIGHT** 1, A form of sight that is inserted into a gun barrel for testing purposes.
2, The act of aligning the axis of the sight with the axis of the bore, which is called 'boresighting'.

**BOULENGE CHRONOGRAPH** An instrument for measuring velocity, invented by Captain-Commandant P.L. le Boulengé of the Belgian Artillery in the 1880s. It consisted of an electromagnet holding a vertical steel bar and connected to a screen of wire. A second screen was connected to an electromagnet controlling a sharp chisel which was alongside the steel bar. To operate, the bar was marked opposite the tip of the chisel and the gun was then fired through the two screens, which were a measured distance apart. As the shot passed through the first screen it interrupted the electric current so that the magnet allowed the steel bar to drop. As the shot passed the second screen it released the second magnet, which caused the chisel to strike the bar and mark it. By measuring the distance between the two marks on the bar, the distance fallen could be calculated in respect to gravity, so giving a very exact measure of the time elapsed between the shot's passage of the two screens, from which it was possible to deduce the shot's velocity. The Boulengé chronograph remained in use as an experimental tool until the perfection of electronic methods of velocity measurement in the 1940s.

**BOURGES ARSENAL** The principal French army gunmaking arsenal. Located in the city of Bourges, about 150 miles south-east of Paris, it was established prior to the French Revolution for the manufacture of cannon. The 75mm M1897 field gun was developed here, though most were built by contractors elsewhere, and with few exceptions all French Army artillery has been developed at Bourges ever since. The most recent products have been the 'TR' 155mm howitzer and the 'GCT' 155mm self-propelled gun.

**A Soviet Army soldier examines a captured German 15cm infantry howitzer with a box trail; it is clear from this photograph that when the gun elevates, the amount of traverse will be restricted by the breech striking the inside of the box.**

**BOX TRAIL** A gun trail made in the form of a box. The advantage is that the gun breech, when the gun is elevated, passes between the sides of the trail and is not restricted as it would be with a pole trail (qv). The drawback is that the traverse of the gun on the carriage is restricted when the gun is elevated by the danger of the recoiling breech striking the sides of the trail.

**BREECH BUSH** A metal ring that screws into the breech ring (qv) of a screw-breech gun, keeps the barrel or liner in place and also has threads cut in it for the reception of the breechblock. By removing the breech bush it is possible to withdraw the barrel liner or 'A' Tube to the rear.

**BREECH MECHANISMS** The apparatus for opening and closing the rear end of the gun to permit breech loading.

There have been innumerable designs of breech mechanisms since the introduction of breech-loading in the mid 19th century; many of the more important are mentioned elsewhere under their individual names. In broad terms, however, there are two ways to close a breech, either by screwing a block of steel into threads prepared in the breech of the gun or by sliding a square block of steel across the rear of the chamber.

The 'interrupted screw' breech uses a cylinder of steel which has a screw thread cut on its outer surface; this thread is then partly milled away to leave threaded segments. The breech ring is then similarly threaded and milled so that the block can be inserted into the ring by sliding the threaded portions of the block through the milled segments of the ring. Once the block has been fully inserted, it can be turned so that the threaded portions engage. In order to offer a greater bearing surface, the 'stepped thread' is universally employed, in which the threaded segments are cut on two diameters and cover two-thirds of the block, while the plain segments occupy the remaining third. The block can then be inserted with the major threaded arcs in the cut-out sections of the breech ring, the lesser threaded arcs passing in beneath the major threaded section of the ring. One-third of a turn then locks the screw in place. Modern breechblocks are so designed that one-sixth of a turn is all that is required to lock them.

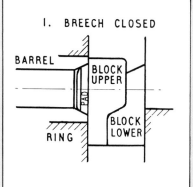

## I. BREECH CLOSED

## 2. PAD WITHDRAWN

## 3. BREECH OPEN

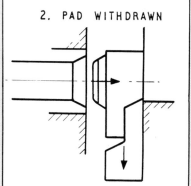

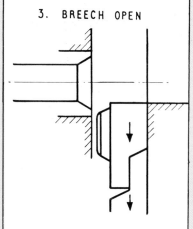

**The most recent innovation is this 'split block' breech developed in Britain in 1980. The design combines the efficiency of the de Bange system of obturation with the simplicity of a sliding block breech**

The sliding block (or 'sliding wedge') breech uses a block of steel that slides in a mortise cut in the breech ring and is so designed that the back of the block bears against the rear of the ring or so that ribs on the side of the block engage in grooves in the side of the ring. When slid open the chamber is exposed for loading; the closing movement, in addition to sliding the block across, also allows it to move slightly forward to press against the rear of the cartridge case or close tightly against the rear of the chamber if a bagged charge is being used.

A modern type of sliding block breech, for use with bag-charge guns, has been designed by the British Royal Armaments Research and Design Establishment for use in a new 120mm tank gun and also in a new 155mm howitzer. This uses a two-piece block, the front portion of which is fitted with a de Bange obturation paid. In operation the front block slides across until the pad is aligned with the chamber mouth, after which the second, wedge-shaped portion slides in behind the first piece and so thrusts the obturation pad into the mouth of the chamber. This design marries the advantages of the de Bange system of obturation with the simple manufacturing and operation of a sliding block breech.

Breech mechanisms may be hand- or power-operated, and may be semi-automatic. A semi-automatic breech opens automatically at the end of the recoil stroke or during the counter-recoil movement and ejects the empty case so that the breech is open and ready for loading as soon as the gun has returned to battery. On loading, the block closes automatically. The use of a semi-automatic breech can reduce the number of men required to serve the gun by making it unnecessary to have one man doing nothing but open and close the breech. Semi-automatic breech mechanisms are now being adopted on bag-charge guns to allow a rapid burst of fire.

**A most unusual breech mechanism on a French 155mm howitzer of 1904. The breech screw rides on two rods that slide beneath the gun. On recoil the rods are held by a catch and as the gun runs out so the breech is opened. A shell and cartridge are placed between the block and the chamber (as seen here) and the breech closed by turning the wheel at the right side, winding the rods back beneath the gun.**

**American 'translating tray' mechanism of the 1890s. The revolving crank turned the screw to unlock it. The 'translating crank' wound the screw on to the 'translating tray'. The 'tray latch' was then released to allow both tray and screw to be swung aside to permit loading.**

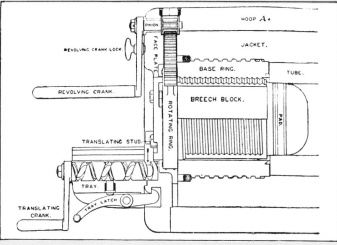

**BREECH RING** A heavy block of steel that surrounds the rear end of the gun barrel and carries the breech mechanism; not necessarily ring-shaped. In addition, it acts as an anchor for the various components of a built-up gun and, to some extent, as a counter-weight to the muzzle preponderance. In the case of breech rings carrying a sliding block, the ring can be 'open jawed' or 'tied jaw'; in the former case the rear face of the block is entirely visible and slides between the two sides of the ring, while in the case of the 'tied jaw' the ring surrounds the block for most of its length, leaving merely an aperture through which the ammunition can be loaded, and the block is therefore not entirely visible.

**BRITISH ARTILLERY** British artillery was almost all built at Woolwich Arsenal until the mid 19th century, though manufacture of some smaller cannon was often contracted out to commercial iron foundries. With the adoption of the Armstrong gun (qv), a large portion of the order was made by Armstrong's Elswick Ordnance Works as well as

**Right: An open-jaw breech ring on a Soviet 45mm anti-tank gun**

by Woolwich, but when manufacture of the Armstrong breech-loader stopped, manufacture of rifled muzzle-loaders once more reverted to Woolwich. But the increased demand for naval ordnance in the latter part of the 19th century, plus orders for coast defence artillery and, in the South African

War, increased requirements for field pieces, all led to an increase in manufacture outside the Government arsenal, and Armstrong, Whitworth, Vickers, the Coventry Ordnance Works and Beardmore all began manufacturing, while Armstrong and Vickers also began designing guns to meet specific

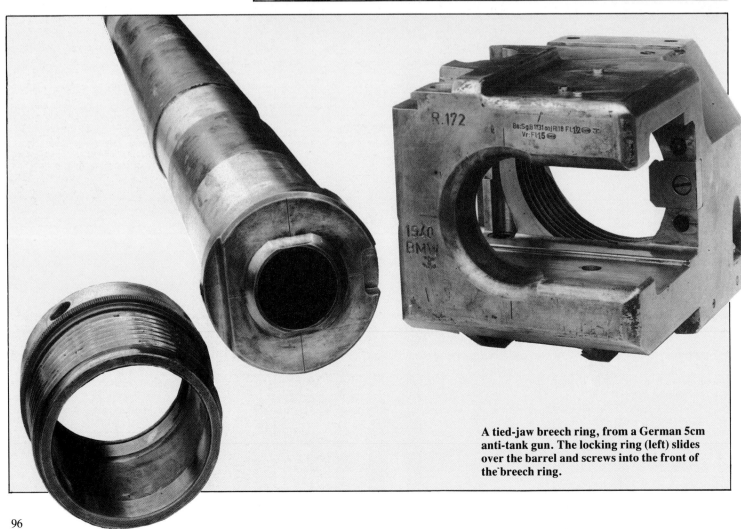

**A tied-jaw breech ring, from a German 5cm anti-tank gun. The locking ring (left) slides over the barrel and screws into the front of the breech ring.**

**Above: Early anti-aircraft gunners with their 13pr 9cwt high-angle gun on Peerless Motor Lorry, France 1917.**

Woolwich requirements. This system continued, to a greater or lesser degree, until 1918. After that however, the Royal Ordnance Factory organization took more and more of the gun manufacturing, and except for major-calibre naval guns the private gunmakers did little. The Ordnance Factory organization expanded further during the Second World War. Vickers-Armstrong continued to manufacture guns, but their output was minor compared to that of the ROFs.

Almost all gunmaking in Britain has been performed at the Royal Ordnance Factory, Nottingham, since 1945 with some small quantities being made on contract by Vickers.

After the Armstrong gun had gone into service, it was largely superseded by rifled muzzle-loading equipments, since these allowed the firing of far heavier charges such as were necessary to overcome warship armour. Moreover, they were of simpler construction and were easier to operate and maintain. But the gradual search for more power led to longer guns. These became difficult to load at the muzzle, and so in 1881 breech-loading reappeared in British service. The introduction of the French M1897 gun, with fixed cased ammunition, on-carriage recoil system, shield and other advances meant that Britain had been overtaken in field gun design, and when the South African War broke out it was necessary to purchase a number of 15pr QF guns from Ehrhardt of Germany in order to bring the field artillery up to strength with a new design. A new field gun, the 18pr, appeared in 1904 together with the 13pr Royal Horse Artillery gun; these were amalgamations of the best features of designs prepared by Vickers, Armstrong and Woolwich Arsenal, and they remained in service until the 1940s. Once the problem of on-carriage recoil was overcome, new designs came in some profusion prior to 1914 and more followed, principally heavy howitzers and guns, but also the first anti-aircraft guns.

## BRITISH ARTILLERY EQUIPMENT

| Equipment | Date | Calibre (mm) | Barrel length (cals) | Breech mech. | Elevation max (deg) | Traverse (deg) | Weight in action (kg) | Shell weight (kg) | Muzzle velocity (m/sec) | Range, max (metres) |
|---|---|---|---|---|---|---|---|---|---|---|
| **Anti-tank artillery** | | | | | | | | | | |
| 2pr Mk 9 | 1936 | 40 | 50 | VSB/SA | 15 | 360 | 796 | | 807 | 5900 |
| 6pr 7cwt Mk 2 | 1941 | 57 | 43 | VSB/SA | 15 | 90 | 1143 | 2.7 | 820 | 5030 |
| 17pr Mk 1 | 1942 | 76.2 | 55 | VSB/SA | 16.5 | 60 | 2097 | 7.7 | 884 | 9145 |
| | | | | | | | | | | |
| **Anti-aircraft artillery** | | | | | | | | | | |
| 2pr Mk 8 Twin | 1939 | 40 | 40 | VSB/A | 80 | 360 | 7570 | 0.9 | 693 | 4875 ceiling |
| 6pr 6cwt Mk 1 | 1944 | 40 | 56 | VSB/SA | 85 | 360 | 11176 | 2.7 | 945 | 10360 ceiling |
| 10pr 'Russian' Mk 1 | 1915 | 75 | 50 | SQF | 90 | 360 | | 4.8 | 825 | |
| 3in 5cwt Mk 1 | 1915 | 76.2 | 23 | SQF | 79 | 360 | 915 | 5.7 | 500 | |
| 3in 20cwt Mk 1 | 1914 | 76.2 | 45 | VSB/SA | 90 | 360 | 2720 | 7.2 | 762 | 11330 ceiling |
| 12pr 12cwt Mk 1 | 1915 | 76.2 | 40 | SQF | 85 | 360 | | 5.7 | 670 | 6100 ceiling |
| 13pr 9cwt AA | 1915 | 76.2 | 32 | SQF | 80 | 360 | | 5.7 | 655 | 5790 ceiling |
| 18pr Mk 2 | 1916 | 84 | 28 | SQF | 80 | 360 | | 8.4 | 492 | 5485 ceiling |
| 3.7in Mk 1 | 1937 | 94 | 50 | HSB/SA | 80 | 360 | 9317 | 12.7 | 792 | 12500 ceiling |
| 3.7in Mk 6 | 1942 | 94 | 65 | HSB/SA | 80 | 360 | 17400 | 12.7 | 1057 | 18075 ceiling |
| 4in Mk 5 | 1915 | 101 | 45 | HSB/SA | 80 | 360 | 6858 | 14 | 716 | 9140 ceiling |
| 4.5in Mk 2 | 1936 | 114 | 45 | HSB/SA | 80 | 360 | 14986 | 24.7 | 731 | 12980 ceiling |
| 5.25in Mk 2 | 1942 | 133 | 50 | HSB/SA | 70 | 360 | 30786 | 36.3 | 853 | 16950 ceiling |

Left: The 3in 20cwt anti-aircraft gun dated from 1914 but remained in use during the first part of World War Two since it was light and manoeuvrable for field army use.

Loading a 4.5in anti-aircraft gun in the London Docks, 1940.

In the 1920s new designs were prepared to replace the World War One designs, but the financial stringency of the times ensured that nothing was actually done until about 1936, when the 25pr gun appeared, followed by the 3.7in anti-aircraft gun. The threat of war speeded up design, but the shortage of manufacturing facilities meant that many new designs were not to appear until 1941, and this was a spur to the expansion of the Ordnance Factory gunmaking facilities during the war.

The wartime designs continued in service until the 1950s, the Korean War being fought entirely with World War Two guns. In the mid-1950s NATO standardized on 105mm for close support and 155mm for general support weapons, and as neither of these had ever been British calibres it meant a complete change of equipment. In order to economize, the first 105mm weapon adopted was the Italian M56 pack howitzer, while the self-propelled 155mm M44 howitzer from the USA became the new general support weapon. The Abbot SP 105mm gun was adopted in the early 1960s, followed by the American 155mm SP M109. Then came the 105mm Light Gun, the most recent British design to be put into service. Other weapons in current use are entirely American, including the 8in (203mm) howitzer, the 175mm gun (obsolescent) and improved models of the 155mm M109. The towed general support gun is the Anglo-German-Italian FH70 howitzer, and a self-propelled version of this was under development for over ten years; the development was abandoned early in 1987 and a less complex and less expensive alternative will be adopted. It is probable that a 155mm turreted gun, designed by Vickers, may be adopted as the future British SP equipment.

**Left: The 25pr gun in final form, fitted with a muzzle brake. It was the standard British divisional field gun from 1940 to 1967 and is still in use in other parts of the world.**

**Right: A 15in howitzer in France, 1916.**

**Below: Gunners of the Allied Mobile Force in Norway with a 105mm light gun, towed by a Volvo over-snow tractor, 1980.**

| Equipment | Date | Calibre (mm) | Barrel Length (cals) | Breech mech. | Elevation max (deg) | Traverse (deg) | Weight in action (kg) | Shell weight (kg) | Muzzle velocity (m/sec) | Range max (metres) |
|---|---|---|---|---|---|---|---|---|---|---|
| *British Artillery Equipment (Cont)* | | | | | | | | | | |
| **Mountain artillery** | | | | | | | | | | |
| 10pr Jointed Mk 1 | 1901 | 69 | 27 | IS | 25 | Nil | 400 | 4.5 | 392 | 5480 |
| 2.75in Mk 1 | 1912 | 69 | 27 | IS | 22 | 8 | 585 | 5.6 | 381 | 6400 |
| 3.7in How Mk 1 | 1915 | 94 | 12 | SQF | 40 | 40 | 730 | 9 | 296 | 5395 |
| | | | | | | | | | | |
| **Field artillery** | | | | | | | | | | |
| Smith Gun | 1941 | 76.2 | 18 | SQF | 40 | 360 | 274 | 3.6 | 122 | 500 |
| 12pr 6cwt Mk 4 | 1900 | 76.2 | 22 | SQF | 16 | Nil | 910 | 5.7 | 483 | 5485 |
| 13pr Mk 1 | 1904 | 76.2 | 24 | SQF | 16 | 8 | 1014 | 5.7 | 510 | 5390 |
| 15pr MK 1 | 1901 | 76.2 | 30 | SQF | 16 | 6 | 1030 | 6.4 | 760 | 5850 |
| 18pr Mk 1 | 1904 | 84 | 28 | SQF | 16 | 8 | 1279 | 8.4 | 492 | 5965 |
| 25pr Mk 2 | 1940 | 87 | 27 | VSB | 40 | 8 | 1800 | 11.3 | 518 | 12250 |
| 25pr Short Mk 1 | 1944 | 87 | 14 | VSB | 40 | 8 | 1365 | 11.3 | 390 | 9875 |
| 105mm Pack L10A1 | 1956 | 105 | 20 | VSB/SA | 65 | 56 | 1273 | 15 | 420 | 9997 |
| 105mm L13 'Abbot' | 1965 | 105 | 37 | VSB/SA | 70 | 360 | 17464 | 16.1 | 705 | 17300 |
| 105mm Light Gun L118 | 1974 | 105 | 30 | VSB/SA | 70 | 11 | 1818 | 16.1 | 617 | 15070 |
| 4.5in How Mk 1 | 1904 | 114 | 13 | HSB | 45 | 6 | 1365 | 15.9 | 307 | 6675 |
| 4.7in Gun Mk 1 | 1895 | 120 | 40 | SQF | 20 | Nil | 3818 | 21 | 655 | 9150 |

The 5.5in medium gun; the two 'horns' alongside the barrel are spring balancing presses which counter the weight of the barrel.

| *British Artillery Equipment (Cont)* Equipment | Date | Calibre (mm) | Barrel Length (cals) | Breech mech. | Elevation max (deg) | Traverse (deg) | Weight in action (kg) | Shell weight (kg) | Muzzle velocity (m/sec) | Range max (metres) |
|---|---|---|---|---|---|---|---|---|---|---|
| **Medium and heavy artillery** | | | | | | | | | | |
| 4.5in Gun Mk 2 | 1939 | 114 | 41 | IS | 45 | 60 | 5842 | 24.9 | 685 | 18745 |
| 60pr Gun Mk 1 | 1904 | 127 | 32 | IS | 21 | 8 | 4470 | 27.2 | 633 | 12700 |
| 60pr Gun Mk 2 | 1918 | 127 | 37 | IS | 35 | 8 | 5464 | 27.2 | 647 | 15000 |
| 5in How Mk 1 | 1895 | 127 | 8 | IS | 45 | Nil | 1212 | 22.7 | 240 | 4390 |
| 5.5in Gun Mk 3 | 1941 | 140 | 30 | IS | 45 | 60 | 6190 | 45.4 | 510 | 14810 |
| 6in Gun Mk 19 | 1916 | 152 | 35 | IS | 38 | 8 | 9376 | 45.4 | 716 | 17145 |
| 6in 26cwt How Mk 1 | 1915 | 152 | 13 | IS | 45 | 8 | 3083 | 45.4 | 376 | 8686 |
| 6in 30cwt How Mk 1 | 1896 | 152 | 14 | IS | 35 | Nil | 3505 | 53.7 | 236 | 4755 |
| 155mm How FH70 | 1975 | 155 | 38 | VSB/SA | 70 | 56 | 9300 | 43.5 | 880 | 24000 |
| 7.2in How Mk 1 | 1941 | 183 | 22 | IS | 45 | 8 | 10323 | 91.6 | 518 | 15450 |
| 7.2in How Mk 6 | 1943 | 183 | 33 | IS | 63 | 60 | 14770 | 91.6 | 586 | 17925 |
| 8in How Mk 1 | 1915 | 203 | 14 | IS | 45 | Nil | 14200 | 90.7 | 396 | 9600 |
| 8in How Mk 7 | 1917 | 203 | 17 | IS | 45 | 8 | 8990 | 90.7 | 457 | 11250 |
| 9.2in How Mk 1 | 1914 | 234 | 13 | IS | 55 | 60 | 13590 | 131.5 | 361 | 9200 |
| 9.2in How Mk 2 | 1916 | 234 | 17 | IS | 50 | 60 | 16460 | 131.5 | 408 | 12750 |
| 9.45in How Mk 1 | 1904 | 240 | 8 | IS | 63 | 19 | 8687 | 127 | 282 | 6995 |
| 12in How Mk 2 | 1916 | 305 | 13 | IS | 65 | 60 | 37190 | 340 | 364 | 10370 |
| 12in How Mk 4 | 1917 | 305 | 17 | IS | 65 | 60 | 38100 | 340 | 447 | 13120 |
| 15in How Mk 1 | 1915 | 351 | 10 | IS | 45 | 25 | | 635 | 340 | 9870 |

8in howitzers being towed by Holt tractors in
France, 1918.

The 60pr (5in) medium gun entered service in
1904 and survived until 1944.

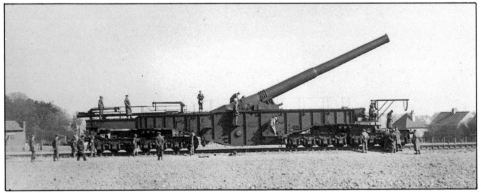

Above: The 6in Mark 7 gun, backbone of the coast defences of the British Empire from the 1880s until 1956.

Left: 'H.M. Gun Scene-Shifter' in a firing position outside Dover in 1941. This 14in gun, manned by Royal Marines, frequently shelled German gun positions in France.

Right: A rush to man 40mm Bofors guns on the South Coast during the 'flying bomb' attacks in the summer of 1944.

*British Artillery Equipment (Cont)*

| Equipment | Date | Calibre (mm) | Barrel Length (cals) | Breech mech. | Elevation max (deg) | Traverse (deg) | Weight in action (kg) | Shell weight (kg) | Muzzle velocity (m/sec) | Range max (metres) |
|---|---|---|---|---|---|---|---|---|---|---|
| **Railway artillery** | | | | | | | | | | |
| 9.2in Gun Mk 3 | 1915 | 234 | 30 | IS | 35 | 20 | 60960 | 172 | 640 | |
| 9.2in Gun Mk 10 | 1916 | 234 | 45 | IS | 30 | 360 | 91450 | 172 | 822 | 19200 |
| 12in How Mk 1 | 1916 | 305 | 12 | IS | 65 | 40 | 58780 | 340 | 358 | 10180 |
| 12in How Mk 3 | 1916 | 305 | 17 | IS | 65 | 40 | 61725 | 340 | 448 | 13715 |
| 12in Gun Mk 9 | 1915 | 305 | 40 | IS | 30 | 2 | | 385 | 762 | 29900 |
| 12in Gun Mk 11 | 1918 | 305 | 50 | IS | 45 | 2 | 495050 | 385 | 838 | 34565 |
| 14in Gun Mk 3 | 1916 | 355 | 45 | IS | 40 | 4 | 251985 | 719 | 746 | 31820 |
| 18in How Mk 1 | 1920 | 457 | 35 | IS | 40 | 4 | 354475 | 1134 | 580 | 20755 |

*British Artillery Equipment (Cont)*

| Equipment | Date | Calibre (mm) | Barrel Length (cals) | Breech mech. | Elevation max (deg) | Traverse (deg) | Weight in action (kg) | Shell weight (kg) | Muzzle velocity (m/sec) | Range max (metres) |
|---|---|---|---|---|---|---|---|---|---|---|
| **Coast artillery** | | | | | | | | | | |
| 3pr Nordenfelt Mk 1 | 1889 | 47 | 45 | VSB | 9 | 360 | 1118 | 1.5 | 556 | 6850 |
| 3pr Hotchkiss Mk 1 | 1885 | 47 | 40 | VSB | 9 | 360 | 1120 | 1.5 | 556 | 6850 |
| 6pr Hotchkiss Mk 1 | 1885 | 57 | 40 | VSB | 20 | 360 | 1521 | 2.7 | 540 | 6860 |
| 6pr Nordenfelt Mk 1 | 1885 | 57 | 42 | VSB | 20 | 360 | 1565 | 2.7 | 540 | 6675 |
| 6pr 10cwt Mk 1 Twin | 1937 | 57 | 47 | VSB/SA | 7.5 | 360 | 10038 | 2.7 | 720 | 4710 |
| 12pr 12cwt Mk 1 | 1894 | 76.2 | 40 | SQF | 20 | 360 | 4190 | 5.7 | 688 | 9235 |
| 4in Mk 3 | 1906 | 101 | 40 | SQF | 20 | 360 | 7036 | 11.3 | 687 | 10240 |
| 4in Mk 5 | 1915 | 101 | 45 | HSB/SA | 20 | 360 | 11585 | 14 | 805 | 13530 |
| 4.7in Mk 2 | 1888 | 120 | 40 | SQF | 20 | 360 | | 20.4 | 647 | 11880 |
| 4.7in Mk 5 | 1900 | 120 | 45 | SQF | 20 | 360 | 8815 | 20.4 | 716 | 15090 |
| 6in Mk 7 | 1898 | 152 | 45 | IS | 16 | 360 | 16282 | 45.4 | 760 | 11520 |
| 6in Mk 24 | 1939 | 152 | 45 | IS | 45 | 360 | | 45.4 | 871 | 19850 |
| 7.5in Mk 2 | 1905 | 190 | 50 | IS | 20 | 360 | 51825 | 90.7 | 853 | 19850 |
| 8in Mk 8 | 1942 | 203 | 50 | IS | 70 | 160 | | 116 | 830 | 26700 |
| 9.2in Mk 10 | 1900 | 234 | 45 | IS | 35 | 360 | 127000 | 172 | 860 | 33550 |
| 9.2in High Angle | 1905 | 234 | 32 | IS | 45 | 360 | 69500 | 131 | 625 | 15180 |
| 10in Mk 3 | 1888 | 254 | 34 | IS | 15 | 360 | 132090 | 227 | 621 | 10500 |
| 13.5in Mk 3F | 1892 | 343 | 30 | IS | 15 | 360 | | 567 | 640 | |
| 15in Mk 1 | 1936 | 381 | 42 | IS | 45 | 480 | 378738 | 880 | 817 | 38405 |

**Moving a 60pr field gun by means of a 'Gun Carrier', France, 1917. It was hoped that this would provide a way to move artillery across mud, but it proved too slow.**

*British Artillery Equipment (Cont)*

| Equipment | Date | Calibre (mm) | Barrel Length (cals) | Breech mech. | Elevation max (deg) | Traverse (deg) | Weight in action (kg) | Shell weight (kg) | Muzzle velocity (m/sec) | Range max (metres) |
|---|---|---|---|---|---|---|---|---|---|---|
| **Recoilless artillery** | | | | | | | | | | |
| 3.45in P1 | 1944 | 87 | 13 | SQF | - | - | 25 | 1.7 | 277 | 500 |
| 3.7in Mk 1 | 1944 | 95 | 22 | SQF | 10 | 20 | 170 | 10.2 | 305 | 1825 |
| 95mm Mk 1 | 1945 | 95 | 27 | SQF | 35 | 60 | 1066 | 11.4 | 487 | 9875 |
| 120mm BAT L1 | 1952 | 120 | 33 | VSB | 24 | 40 | 1000 | 12.8 | 462 | 1000 |
| 7.2in P1 | 1944 | 183 | 18 | IS | 15 | 30 | 1625 | 54.4 | 275 | 3110 |

### BRITISH SELF-PROPELLED ARTILLERY

| Equipment | Date | Calibre (mm) | Chassis | Elevation max (deg) | Traverse on mount (deg) | Weight in action (kg) | Range, max (metres) | Ammunition carried (rds) |
|---|---|---|---|---|---|---|---|---|
| Abbot | 1964 | 105 | Special | 70 | 360 | 16560 | 17000 | 40 |
| Achilles | 1944 | 76 | M10 SP | 20 | 360 | 29580 | 2000 (A/Tk) | 50 |
| Archer | 1944 | 76 | Valentine | 15 | 45 | 16255 | 2000 (A/Tk) | 52 |
| Birch Gun | 1928 | 84 | Vickers 'C' | 37.5 | 360 | 12655 | 9600 | ? |
| Bishop | 1942 | 87 | Valentine | 15 | 8 | 17450 | 5850 | 32 |
| Deacon | 1942 | 57 | AEC truck | 15 | 360 | 12200 | 1000 (A/Tk) | 24 |
| Sexton | 1943 | 87 | Ram | 40 | 50 | 25860 | 12255 | 112 |
| Vickers AS90 (Prototype) | 1986 | 155 | Vickers | 70 | 360 | 42200 | 2400 | 50 |

**BROADWELL RING** A method of obturation (qv) used with early breech-loading guns, the name being derived from the inventor. It consisted of a soft metal ring let into the face of a sliding breech-block so that the closing movement of the block, which included a slight forward motion, pressed the ring against a prepared face on the rear of the chamber, so making a gas-tight seal. It was used in the 1875-85 period with early Krupp and other guns that fired bagged charges, but was eventually found to have a limited life and to be unreliable with heavy charges. A much-improved version was under development in Germany in 1945, and the idea was taken and perfected by the British for the 120mm tank gun used in the Chieftain tank.

**BRUCE** A British experimental gun developed during the Second World War. It consisted of a 13.5in gun barrel into which an 8in calibre liner was fitted, extending beyond the muzzle of the 13.5in portion. Two guns were built by Vickers-Armstrong; one was assembled on a proof mounting on the Isle of Grain and fired for experiments, while the other was assembled into a barbette mounting near Dover and also used for long-range trials. The gun used deep 16-groove rifling and fired a shell formed with curved ribs which fitted the grooves. It fired a 116kg (256lb) shell at 1,400m/sec (4,593ft/sec) to a maximum range of 100.5 kilometres (62.5 miles). The gun had a life of about 30 rounds before erosion made the barrel inaccurate. The gun was named after Admiral Sir Bruce Fraser, then Controller of the Navy, since the gun at Dover was manned by Royal Marines. Both guns were scrapped after the war.

**Bruce, the 13.5/8in hyper-velocity gun, during its construction.**

**BRUMMBAR** German assault gun in service 1943-5; more correctly known as the 'Sturmpanzer IV'. It consisted of a 15cm short-barrelled howitzer mounted frontally in a boxy superstructure built on the hull of the Panzer IV tank. Designed in 1942, 60 were ordered by Hitler, the first batch to be ready by May 1943. After this first 60 were built, some small modifications were made and long-term production went ahead, a total of 306 being made before production ceased in March 1945. The howitzer was capable of 10°traverse on either side of centre, elevated to 30° and had a maximum range of about 6 kilometres (4 miles). A total of 38 rounds of ammunition was carried. Brummbär was operated by a five-man crew.

**Left: Brummbär (Grizzly Bear) the German 15cm self-propelled assault gun.**

**BRUNO** 'Family name' for a group of German railway guns of broadly similar characteristics. 'Theodor Bruno' was a 24cm (9.5in) gun weighing 92 tons (93.5 tonnes) and firing a 327lb (148kg) shell to 12.5 miles (20 kilometres). Six were built in 1936-9.

'Kurze Bruno' was a 28cm (11in) gun weighing 126 tons (128 tonnes) and firing a 529lb (240kg) shell to 9 miles (14.3 kilometres). Eight were built in 1937-8.

'Lange Bruno' was another 28cm (11in) gun weighing 121 tons (122.9 tonnes) and firing a 626lb (284kg) shell to 22.4 miles (36 kilometres). Three were built in 1937.

'Schwerer Bruno' was also a 28cm (11in) gun, weighing 116 tons (117.8 tonnes) and firing a 626lb (284kg) shell to 22 miles (35.7km). Only two of these were built, in 1937-8.

'Neue Bruno' was a more modern design (the others were based on ex-naval guns on mountings derived from First World War patterns) manufactured by Krupp in 1940-2. It weighed 147 tons (149.3 tonnes) and fired a 584lb (265kg) shell to 23 miles (36.6km). Three were built.

The 28cm Kurze Bruno railway gun on a turntable and ready for firing.

**BUFFER** A component of a recoil system that absorbs the recoil force during the rearward movement of the gun. The buffer is invariably a cylinder containing oil through which a piston is moved by the action of the recoiling gun. Either the buffer is attached to the mounting and the gun pulls the piston, or the piston-rod is attached to the mounting and the gun moves the cylinder across the piston-head. In either case, there is a valve system which, at the start of movement, is open so that oil can pass from one side of the piston-head to the other, the restriction giving some braking effect. The valve is gradually closed as the recoil movement proceeds, thus creating greater resistance for the oil and thus increasing the braking effect. Eventually the valve closes completely and the gun is brought to rest. The initial setting of the valve can be altered by a cut-off gear (qv) so that the recoil stroke is shortened as the gun elevates and prevents the breech from striking the ground. Alteration of the valve orifice during recoil can be done by rotary motion, given by a helical groove cut into the inside of the cylinder, or by cutting a slot in the piston and allowing a tapering bar, on the inside of the cylinder, to run through the slot and so vary the free space. There are also other, more complicated, systems.

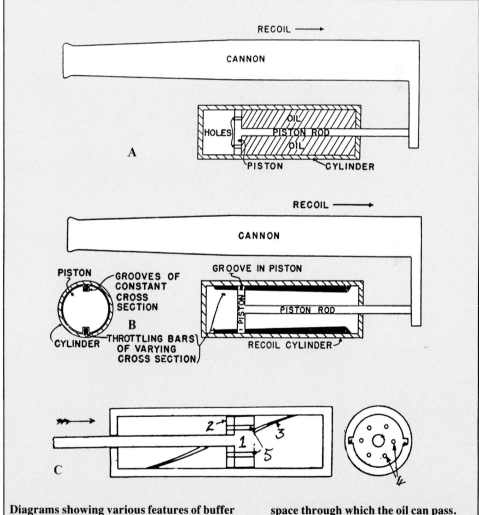

Diagrams showing various features of buffer systems:
(A) A simple hydraulic buffer; the piston is pulled through the oil, some of which passes through the holes in the piston-head and acts as a brake.
(B) Gradual increase of the braking effect can be achieved by using throttling bars which pass through grooves in the piston-head and gradually reduce the space through which the oil can pass.
(C) Another way of increasing the braking effect; the piston-head is in two parts, (1) the head itself and (2) a rotating portion, controlled by lugs which engage in (3) a spiral groove. As the piston moves on recoil, so the action of the groove rotates the valve (4) and gradually closes the holes (5) in the piston-head.

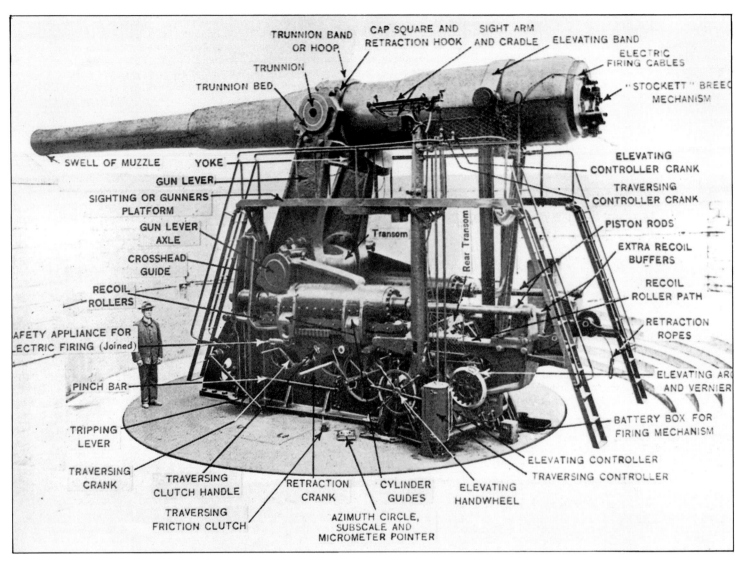

TRUNNION BAND OR HOOP.
CAP SQUARE AND RETRACTION HOOK
SIGHT ARM AND CRADLE
ELEVATING BAND
TRUNNION
ELECTRIC FIRING CABLES
TRUNNION BED
"STOCKETT" BREEC MECHANISM
SWELL OF MUZZLE
YOKE
GUN LEVER
ELEVATING CONTROLLER CRANK
SIGHTING OR GUNNERS PLATFORM
TRAVERSING CONTROLLER CRANK
GUN LEVER AXLE
Transom
Rear Transom
PISTON RODS
CROSSHEAD GUIDE
EXTRA RECOIL BUFFERS
RECOIL ROLLERS
RECOIL ROLLER PATH
AFETY APPLIANCE FOR LECTRIC FIRING (Joined)
RETRACTION ROPES
PINCH BAR
ELEVATING ARC AND VERNIER
TRIPPING LEVER
BATTERY BOX FOR FIRING MECHANISM
TRAVERSING CRANK
TRAVERSING CLUTCH HANDLE
RETRACTION CRANK
CYLINDER GUIDES
ELEVATING HANDWHEEL
ELEVATING CONTROLLER
TRAVERSING CONTROLLER
TRAVERSING FRICTION CLUTCH
AZIMUTH CIRCLE, SUBSCALE AND MICROMETER POINTER

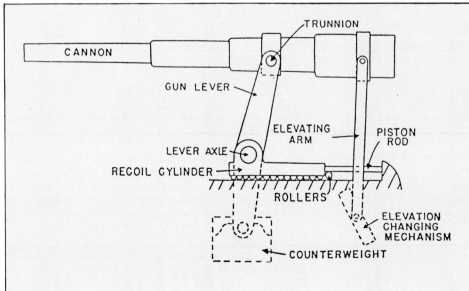

CANNON
TRUNNION
GUN LEVER
ELEVATING ARM
PISTON ROD
LEVER AXLE
RECOIL CYLINDER
ROLLERS
ELEVATION CHANGING MECHANISM
COUNTERWEIGHT

**Above: The US 12in gun M1900 on disappearing carriage M1901, a typical Buffington Crozier design. This illustration from the gun's handbook details all the parts of the mounting.**

**Left: The basic features of the Buffington Crozier disappearing gun mounting. The lever axle and recoil cylinder move back on the rollers as the gun descends and lift the counter-weight.**

## BUFFINGTON-CROZIER MOUNTING

Amercian pattern of disappearing carriage (qv), named after its two inventors, Colonel A.R. Buffington and Captain W. Crozier, both of whom subsequently became Chiefs of Ordnance, US Army. This mounting differs from other types by having the gun arms pivoted on a sub-carriage that can move on an inclined plane, controlled by hydraulic buffers. The upper ends of the gun arms carry the gun, the lower ends being connected to a counter-weight which, in the case of the heaviest guns, could weigh up to 150 tons. This counter-weight moved in a pit beneath the gun mounting. On firing, recoil drives the gun back and down forcing the top end of the gun arms to describe an arc. As this happens, the pivot moves back along the inclined plane, while the lower ends of the gun arms lift the counter-weight. At the end of the recoil movement, the gun has described a somewhat complex path which, in effect, first withdraws the muzzle behind the parapet and then lowers it into the emplacement. The gun comes to rest close to the emplacement floor and is held there by a ratchet operating on the counter-weight. After the gun has been loaded, the ratchet is released and the counter-weight pulls the gun back up to the firing position. The mounting was used with 6, 8, 10, 12, 14 and 16in guns in US Coast Artillery between about 1895 and 1917, after which construction of the type ceased, though guns on this type of mounting remained in use throughout World War Two. Specimens can still be seen at Fort Mills (Corregidor) and Ford Casey (Washington State).

**BUILT-UP GUN** A general term for any gun which is constructed by shrinking hoops (qv) on to the barrel (and each other) to build up a compound construction in layers, and arranged so that the maximum number of layers appear at the points of greatest internal strain.

**BURNEY GUNS** These were British recoilless guns developed by Sir Dennis Burney and the Broadway Trust Company during the Second World War. Burney was a very talented engineer and designer who began investigating recoilless guns in 1941. His design used a chamber that had perforated walls and was surrounded by an outer annular chamber. The cartridge was of brass, pierced with large holes, so that when it was exploded, about four-fifths of the gas developed by the explosion went through the holes in the case and in the inner chamber and passed into the outer annular chamber. This latter chamber was provided with ports at its rear end, fitted with venturi jets, through which the high-pressure gas was directed. The gas was accelerated as it passed through the venturi, and the rush of gas to the rear counterbalanced the recoil due to the ejection of the shell from the muzzle of the gun.

The first full-sized Burney gun was of 3.45in calibre and could be fired from a man's shoulder. It was followed by a 3.7in anti-tank gun and a 95mm field howitzer. All three were under trial at the end of the war, and the 95mm was being studied as a possible armament for airborne artillery units. Another design was a 7.2in bag-charge recoilless gun that had been developed as a means of attacking the fortifications of the Atlantic Wall on the invasion of Europe. This, however, was dropped in favour of another type of weapon. Eventually it was

decided to put the whole idea of recoilless guns back for more research, as a result of which Burney's guns were abandoned and, after several years of development, a totally different weapon – the 120mm BAT (qv) – appeared.

Burney also built an 8in recoilless heavy gun, intended to fire a 200lb (91kg) shell to 20 miles (32 kilometres). It used a novel two-cartridge layout, which malfunctioned when the gun was first fired and destroyed the weapon.

The Burney guns were designed to fire a special lightweight shell which Burney invented; this he called the 'Wall-buster', because it had been designed to defeat reinforced concrete. In essence it consisted of a thin shell carrying a charge of plastic explosive and a base fuze. On striking the target the plastic explosive was deposited like a poultice and the base fuze then detonated it, so generating a shock wave in the target which blew large pieces off the inside of the concrete defence. The 'Wall-buster' formed the basis of the later 'Squash-head' anti-tank shell which is in wide use today.

**BUSY LIZZIE** – a German nickname for the 15cm High Pressure Pump (qv).

**The 7.2in Burney gun was earmarked for the invasion of Europe, but it was never used in its intended role.**

# C

**CALIBER BOARD** Popular name for what was properly known as the 'Westervelt Board' (qv), which sat in 1919 to determine the future equipment of United States Army artillery.

**CALIBRATING SIGHT** A gun sight that permits compensation for changes in the muzzle velocity of the gun. A sight that had, say, a scale of ranges in which the required range was set against a fixed marker would only be accurate so long as the gun projected its shell at the muzzle velocity for which the sight was designed. As soon as the velocity changed, as it would from wear, the range relationship would no longer be correct. If the range is marked on a scale by a curve, calculated to compensate for changes in velocity, and the marker is also adjustable against a scale of velocities, then by altering the marker periodically (whenever the gun is calibrated [see below]) and bringing the range curve into alignment, compensation for the changed velocity will be automatically incorporated.

**CALIBRATION** The process of determining the muzzle velocity of a gun in order that guns of similar velocity can be grouped to place their shells close together on the target, and so that guns with calibrating sights (above) can be periodically corrected. Velocity used to be measured in experimental establishments by the Boulengé chronograph (qv) and in the field by firing, under carefully controlled conditions, at a target a precisely measured distance away and by comparing the result with the theoretical result which would have been achieved for a stated velocity. The use of photoelectric 'skyscreens' became standard in the 1940s; the passage of the shell over a pair of photo-electric cells a measured distance apart caused a timing device to operate. The use of radar techniques became universal in the 1970s to track the shell and measure its speed by the Doppler shift of the reflected radar frequency.

**CALIBRE (OR CALIBER)** The diameter of the inside of the gun's bore, defined as the diameter of a cylinder which fits inside the lands of the rifling. Also used as a measure of a gun's length; thus, a 6in 50-calibre gun is 50 × 6 inches = 300 inches long. Calibre-length is a useful yardstick of the relative power of guns of the same calibre, since the longer the gun the more powerful it is likely to be. It is generally indicated by the notation 'L/...', eg L/70, meaning that the barrel is 70 calibres long.

**CANET BREECH MECHANISM** A system of breech closure used by the Schneider company of France on various guns prior to 1939. The breechblock takes the form of a hemisphere, the outer edges of which are cut like a saw-tooth blade to present a number of locking surfaces. The block pivots on an axis across the flattened face of the hemisphere, moving in a semi-circular path cut into the breech ring. The spherical face closes the chamber, while the outer face has a groove which, when the breechblock is rotated to open, acts as a continuation of the floor of the chamber and as a loading tray. The block contains a percussion firing mechanism, and can be arranged to open semi-automatically during run-out.

The system appears to have been first used with the 240mm Modèle 1903 gun; another major application was to a 90mm anti-aircraft gun developed in the mid-1930s. One drawback to the design is that the cartridge case must be made with a concave base.

**Below left: A diagram of the Canet breech mechanism.**

**Below: The Canet breech mechanism fitted to an experimental French anti-aircraft gun of the early 1930s.**

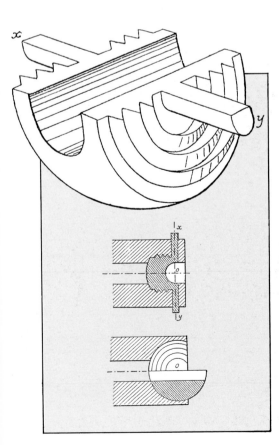

**A cap square retaining the trunnion of a gun in its seat on the top carriage.**

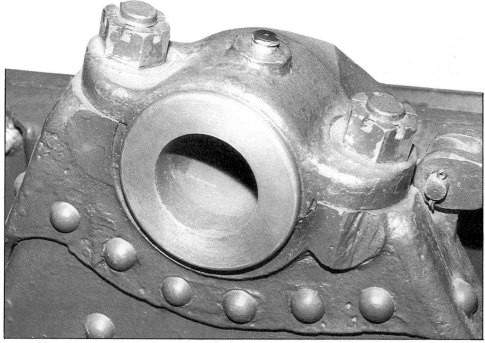

**CAPSQUARE** The curved plate that retains the trunnions of a gun or its cradle in the trunnion bearings on the carriage.

**CARRIER** A hinged arm or unit that supports the breech screw during opening and closing.

**CARRONADE** A short muzzle-loading cannon designed by General Robert Melville and first cast by the Carron Iron Company of Carron, Stirlingshire, Scotland in 1776. It was lighter than ordinary guns of the time and fired a heavier shot to a range of about 500 yards. A significant feature was that the shot and the bore were closely matched to reduce the amount of 'windage' (the difference between the two) and thus confine the power of the propellant gas so that its effect was not wasted by leaking past the shot, as was common with much of the artillery of the time. Calibres up to 10 inches were developed, and it was principally used as a naval gun because it was well-suited to the short-range tactics of the Royal Navy of the time. A small number was to be found in fortifications, as close-defence guns, but these carronades were never adopted as field pieces due to their short range.

**A light carronade on its special carriage; principally a naval weapon, it was generally too short ranged for use in land warfare, although a few were used in fortresses.**

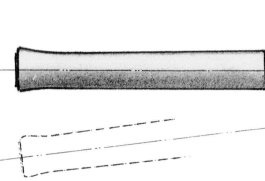

**CASCABLE** The ring or button in the rear centre of the breech end of a muzzle-loading gun. It was of use during the manufacture of the gun as a means of lifting it or centering it in machines. In service, the cascable formed a purchase for the 'breeching rope', which was passed round the gun and used for hauling it back into the firing position after it had recoiled; it was also useful as a purchase for a rope used to swing the gun from side to side; as a bearing for a form of elevating screw; as a visual marker to tell the gunlayer where the axis of the gun was; and as a scale, marked in degrees, to give the gunlayer some idea of the amount of elevation he was applying.

**CASE OF LAYING** A term used with coast defence, fortress and anti-aircraft guns, denoting the method of applying the gun data and laying the gun. *Case 1*: the gun is laid on the target for both line and elevation by the sight. *Case 2*: the gun is laid for line (pointed) by means of the sight and laid for elevation (range) by means of an elevation scale or follow-the-pointer dial (qv). *Case 3*: the gun is laid for both line and elevation by scales or follow-the-pointer dials.

**CEILING** The maximum vertical range achieved by an anti-aircraft gun. There are three types of ceiling:
1, *Maximum ceiling*, the height to which the gun will send the shell if it is elevated to 90°, ignoring any fuze setting. An impressive but impractical figure, as few guns can elevate to 90°.
2, *Effective ceiling*, the height to which a gun will fire a shell when laid at its maximum elevation and with the fuze at its maximum time setting. This, while practicable, is deceptive since it applies to one small area of the sky, and as soon as the gun moves in elevation to track a moving aircraft the effective ceiling no longer obtains.

**A 32pr smoothbore muzzle-loader, showing the cascable at the rear end.**

3, *Practical ceiling*, the greatest height at which it is possible to engage a moving aircraft for a reasonable period of time, sufficient to offer a chance of hitting. This varies according to who is defining the time interval and when. One British definition was 'that altitude at which the gun can engage an approaching target for 20 seconds before the elevation reaches 70°'.

An example of these three figures: when the 3.7in AA gun was introduced into the British Army in 1936 its maximum ceiling was 41,000 feet (12,496 metres), its effective ceiling was 28,000 feet (8,534 metres) and its practical ceiling was 23,000 feet (7,010 metres). Improvements in fuzing later lifted this last figure to 32,500 feet (9,906 metres). It will be appreciated that the development and adoption of proximity fuzes, which have no time limitation, changed the concept of the effective ceiling and virtually equated it with the maximum ceiling, but had little effect on the practical value.

**CENTRAL PIVOT** A type of gun mounting used with fixed artillery in fortresses or coast defence. A cylindrical drum carries a racer (qv) on its top surface, and on this the gun carriage rotates on rollers. The carriage is located by a heavy steel pivot which is central to the drum and engages in a suitable collar in the gun mounting. The mounting carries a platform upon which the gunners can stand and move to load and service the gun. This platform is surrounded by a concrete floor at the same level in which there are lifts to deliver the ammunition from the magazines. The front of the floor and the mounting are protected by a parapet, over which the gun muzzle protrudes. Known in American terminology as 'center pintle'.

**CENTRIFUGAL CASTING** A method of manufacturing gun barrels by spinning the mould as the molten steel is poured in. Centrifugal action causes the metal to take the shape of the mould before it solidifies, and it also gives a very high-quality casting. The method is faster than the conventional method of forging a solid billet.

**CHAMBER** The part of the gun bore in which the cartridge is exploded. In early guns it was merely part of the bore, and of the same diameter as the rest, but in the 19th century it was discovered that making the chamber larger than the calibre gave ballistic advantages. Guns so built were known as 'chambered' guns, and it was necessary to stop the rifling so that the muzzle-loaded shell was not pushed all the way back into the chamber and there was a constant volume in which the charge could explode. With the arrival of breech-loading, the enlarged chamber became normal, though with guns using cased cartridges there is rather less scope for enlarging the chamber in all dimensions and the enlargement has to be in length, with the cartridge case designed to suit. Thus in many cases the propelling charge does not take up all the space in the cartridge.

**CHASE** The portion of a gun which is in front of the trunnions. 'Chase sights' were sights fitted to the chase of heavy coast defence guns mounted behind iron shields; these guns were so large that the line of sight across the breech was obscured by the upper edge of the shield, and the chase sight carried a small mirror behind the rear sight so that the gun could be aimed from the side. 'Chase-hooping' was a method of strengthening guns by shrinking an additional cylinder of steel over the chase.

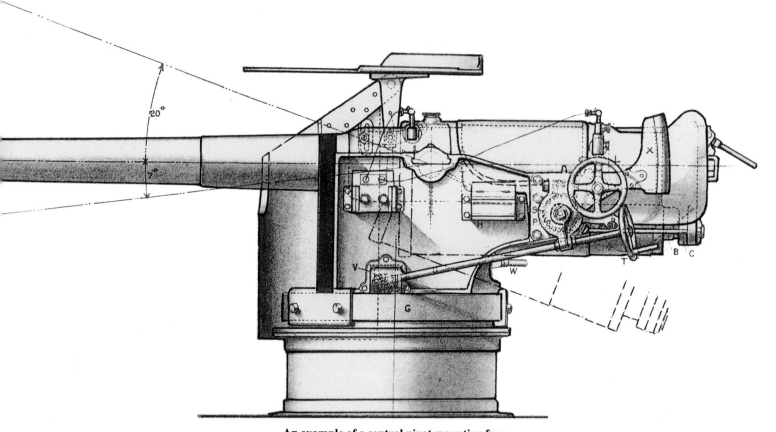

**An example of a central pivot mounting for a coast defence gun.**

**CHINESE ARTILLERY** Chinese artillery in the breech-loading era was entirely purchased from European manufacturers, notably Krupp and Schneider, augmented by weapons captured from Japan, until the emergence of the Communist state in 1949. After this the Nationalist force (ie, Taiwan) adopted American equipment, while the People's Republic adopted Soviet weapons. A number of local designs have been developed since the 1960s; these have generally been local modifications of Soviet weapons. Thus the 'Type 83' gun-howitzer is actually a copy of the Soviet 122mm D-30 howitzer and the Type 66 gunhowitzer is the Soviet 152mm D-20. In the late 1970s entirely new designs, notably of self-propelled equipments, began to appear and Chinese weapons are now being exported to countries in the Middle and Far East. The most recent announcement has been of the 152mm Type 83 towed gun, which fires to a maximum range of 30,370 metres (32,900 yards), and its self-propelled equivalent the Type 83 SP Howitzer, with similar performance. Late in 1985 it was revealed that the '155mm Type WAC 21 Gun-Howitzer' had been developed; this appears to be identical to the Austrian Noricum G-45 and is derived from the S.R.C. (qv) design, using extended-range full bore projectiles. The adoption of 155mm calibre in a country that has hitherto only used 152mm is a significant step, indicating a move away from the restrictive Soviet calibres and also an awareness of international marketing demands. Full details of recent Chinese designs are not yet available in the West.

| CHINESE SELF-PROPELLED ARTILLERY | | | | | | | | |
|---|---|---|---|---|---|---|---|---|
| Equipment | Date | Calibre (mm) | Chassis | Elevation max (deg) | Traverse on mount (deg) | Weight in action (kg) | Range, max (metres) | Ammunition carried (rds) |
| Type 54-1 | 1975? | 122 | Special | 63 | 45 | 15300 | 11800 | 40 |
| Type 83 | 1983 | 152 | Special | 65 | 60 | 30000 | 17230 | 30 |

**Above: An experimental 120mm self-propelled anti-aircraft gun designed by Walter Christie in 1920 for the US Army.**

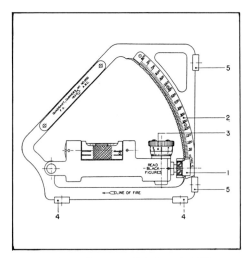

**An American 'Gunners Quadrant'. 1 Clip which locks the arm carrying the level bubble. 2 Elevation scale 3 Fine adjustment. 4 and 5 Feet for resting on the prepared plane on the gun.**

**CHRISTIE,** J. Walter Christie (1866-1944) was an American inventor and engineer who was involved in the design and construction of heavy naval gun mountings in the early years of the 20th century. Shortly before the First World War he went into the automobile business and founded the Front Drive Company, which made military trucks during the war. Christie then revived his interest in artillery and designed self-propelled tracked mountings for the 8in howitzer and, later, for a 4.7in anti-aircraft gun. Neither was particularly successful, and he rapidly lost interest in this field and went on to design a number of high-speed tanks.

**CLINOMETER** Broadly, an instrument for measuring slopes; specifically, in the artillery context, an instrument for measuring the angle of elevation of the gun barrel. It usually takes the form of an adjustable spirit-level attached to the elevating portion of the gun. The clinometer is set, by a scale, to the desired elevation, thus throwing the bubble out of level; the gun is then elevated until the bubble is level, whereupon the gun is at the desired angle of elevation. It may also be a separate instrument that can be placed on a prepared flat surface machined into the gun, to check the accuracy of the built-in clinometer of the sight unit or to apply an angle of elevation to a gun which has sights graduated only in range. Known in American service as a 'Quadrant'.

A German coast defence battery armed with a 15cm L/45 gun; Belgium.

**COAST DEFENCE ARTILLERY** A class of artillery designed to combat warships attempting to raid or bombard shore installations. Coast defence artillery has existed from the earliest times, but it only became a serious branch of artillery in the mid-19th century with the advent of armoured ships the combat of which demanded heavy and accurate weapons capable of breaching their armour at long ranges, outside the range at which the ship could bring accurate fire to bear on the land. The shore guns required protection against the warships'

guns, so coast artillery gun design became linked with developments in fortification. The first major coast defences used heavy rifled muzzle-loading guns in 'casemates' (arched rooms set within a fort constructed of masonry), the rooms having armour-plate shields on their outer wall with ports through which the guns could fire. As guns became more powerful the casemate became too restrictive; moreover, experience in the American Civil War showed that casemates made good targets and that guns in the open, protected by earth and concrete,

were more likely to survive. At about the same time the invention of the disappearing carriage meant that guns could be emplaced in pits in the open and their presence only revealed when they fired, and then only for a few seconds, so that retaliatory fire from ships was almost impossible. The disappearing gun became the major coast defence weapon until the early years of the 20th century, when it was appreciated that with the increased engagement ranges due to advances in gunnery even a gun mounted in a relatively exposed position was an extremely small target and unlikely to be struck by fire from a warship. This led to the adoption of barbette (qv) emplacements, where the gun mechanism was largely protected by a concrete emplacement but the shielded barrel was permanently exposed above the parapet, and this system remained standard thereafter. During the Second World War the advent of air power meant that open gun batteries were now vulnerable to air attack; overhead protection was built in some cases, in others the casemate was reverted to, though in an improved form. Some countries used armoured turrets, similar to those employed on ships, to protect guns which might otherwise be in particularly exposed positions.

Other types of coast artillery included light quick-firing guns of 57-120mm calibre for the close defence of harbours against attacks by torpedo-boats and similar light craft, and heavy howitzers that were sited behind crests, invisible from the sea, so that they could pitch armour-piercing shells high into the air to drop in a steep trajectory and thus pierce the decks of warships. Coast artillery batteries were often also responsible for defending stretches of water by means of submarine mines controlled by electricity and by the use of land-mounted torpedo tubes.

Fire control became a major concern. The warship was a moving target, and some extremely clever instruments were devised to locate and track the target, calculate gun data and transmit information to the guns. An advantage that the coast defence organization enjoyed was almost unlimited space for deploying observation posts and rangefinding apparatus, and for accommodating large mechanical calculators for working out gun data. Radar was incorporated into coast artillery fire control in the early 1940s.

Coast artillery was generally abandoned in the period 1946-56, the view being taken that modern developments of guided missiles and nuclear weapons rendered such installations too easily circumvented or destroyed. Some countries, however, retained their coast defences – Norway, Spain, Sweden, Yugoslavia, and Turkey, to name but a few – and at the time of writing there are signs that other countries are beginning to revive their coast artillery, though often using missiles rather than guns.

Two 9.2in guns sited in the turrets of a British coast defence battery.

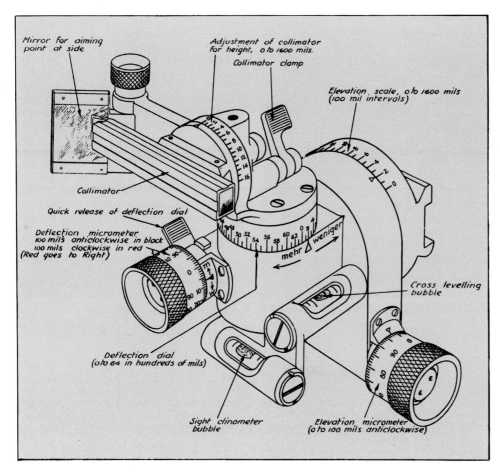

Above: A typical collimating sight, as used with a German World War Two mortar.

Below: This rear view of a rifled muzzle-loading gun on a garrison carriage and slide shows the compressor plates between the side members of the slide.

**COCKERILL** Belgian engineering and gunmaking company, founded in 1817 by John Cockerill, an Englishman. The factory was established at Seraing and was the first Belgian firm to produce steel by the Bessemer process. By 1914 it was comparable with Krupp's for its size and range of products, but after the war the company declined in importance. The company still exists as a subsidiary of a larger firm, Mecar, and manufactures a 90mm tank and anti-tank gun which is in wide use. Cockerill manufactured a range of field guns in the period 1890-1914 and these were used by the Belgian and other armies.

**COLD-WORKING** An American expression that means the same as 'Auto-frettaging', submitting a gun to intense internal hydraulic pressure in order to strengthen it. See 'Auto-Frettage'.

**COLLIMATING SIGHT** A form of quasi-optical sight in which the gunlayer keeps both eyes open, using one to view the target or aiming point and the other to view a marker – usually a cross or triangle – projected at infinity in a sighting unit. The result is that he sees the marker superimposed on the target, and by moving the gun brings the two into alignment. Normally a collimating sight is only used with low-power short-range weapons such as mortars.

**COMPRESSOR** A form of recoil control used with fortress and coast defence guns in the 1860-80 period. The gun was mounted on a carriage that recoiled along a slide. Hanging from the carriage were a number of iron plates, arranged parallel with the gun's movement. Set into the slide were a similar set of plates that could be squeezed together by means of a screw jack. The carriage plates ran between the slide plates, and by tightening the jack the amount of friction could be controlled, to regulate the gun and carriage to recoil and stop at a convenient place for muzzle-loading. The compressor screws were released after loading so that the carriage could be run forward into the firing position, and the compressor was then tightened up again before firing. An alternative system, used in the USA during the American Civil War, had two plates hinged to the side of the carriage which were squeezed against the sides of the slide by means of screw jacks.

**COMPUTING SIGHT** A type of sight used with light anti-aircraft guns. The sight is adjusted at the commencement of the engagement by dialling in values for the target's course, height and speed; the sight then computes the angle and amount of aim-off required and sets the sight graticule accordingly. Some sights will thereafter make an adjustment to compensate for the approach of the target, in others this may have to be adjusted periodically by the gunlayer's assistant. Early computing sights were mechanical, driven by clockwork motors that needed to be rewound after each engagement; modern computing sights use micro-electronic circuitry and can be automatically fed with data from laser rangefinders, radars or other sensors capable of determing the necessary parameters.

**CONCENTRATION** 1, Artillery fire from a number of guns directed against one spot. 2, The act of applying corrections to the azimuth ordered to the target so that each gun's line of fire is directed to the same spot instead of being parallel.

**CONTINUOUS MOTION BREECH** A screw breech mechanism for bag-charge guns in which the motion of rotating the screw, withdrawing it and swinging it clear of the breech to permit loading were performed in sequence by gearing that was driven by the continuous rotation of a handle or by an hydraulic motor.

**COPPER CHOKE** A form of obstruction in a gun barrel as deposits of copper or gilding metal build up due to abrasion from the driving band of the shell. If left to accumulate, this deposit can severely reduce the calibre to the point where it may give sufficient resistance to the shell's movement to cause premature detonation. It can be removed by chemical treatment, but may be prevented by including metal foil in the gun cartridge. This vaporizes in the explosion of the propelling charge and forms a brittle amalgam with the copper deposit, which is then swept away by the passage of the next shell.

**COSMOLINE** a heavy greasy preservative compound used by the US Artillery to coat guns when they are not in use and so preserve them from rust. As the US Coast Artillery only fired their guns once or twice a year in peacetime, for the major part of their life the guns were encased in Cosmoline, and for this reason the coast gunners were, impolitely, called 'Cosmoliners' by the Field Artillery. The coast gunners, in retaliation, called the field gunners 'Redlegs', from the chafing of the skin caused by riding horses or from the red stripe of their dress trousers – opinions differ.

**COUNTER-WEIGHT** A mass of metal placed close to the breech of a gun in order to balance it about the trunnions and obviate the need to fit balancing gear or equilibrators. It is only possible when the gun design is fairly close to balance anyway, or where the additional weight is of little or no consequence; one good example of the former was the addition of a counter-weight over the chamber of the British 25pr gun in order to offset the additional weight at the other end due to the addition of a muzzle brake. The British 3.7in AA gun when on a static mounting also used a counter-weight carried on an arm over the breech, whereas the mobile version used balancing gear as it needed to be light enough to tow behind a vehicle and be put into action by manpower.

**Right: The counter-weight over the chamber of a 25pr gun to compensate for the weight of the muzzle brake.**

A typical continuous motion breech; rotation of the handwheel first unscrewed the block, then withdrew it, and finally rotated the carrier to the open position.

**COURSE AND SPEED SIGHT** An early type of sight for light anti-aircraft guns in which the target's direction of travel and estimated speed were set, thus displacing the sight in relation to the axis of the gun. By pointing the sight at the approaching target, the barrel pointed ahead of it and thus the necessary aim-off was incorporated into the gunlaying process. The sight was not a computing (qv) sight, so it was necessary for one man on the gun detachment to keep altering the values as and when the target's course and speed altered.

**COVENTRY ORDNANCE WORKS** British armament manufacturer created in 1905 by three shipbuilding companies, John Brown, Fairfield Shipbuilding and Cammell-Laird in an endeavour to counter the near-monopoly on naval guns held by Vickers and Armstrong at that time. Its first move into land service artillery came shortly afterwards when it received an order from the War Office to manufacture 18pr field guns. Coventry designed a 4.5in howitzer which succeeded in competitive trials against Vickers and Armstrong designs and which was taken into British service in 1909 and remained there until 1944. In 1910 the company was invited to design a 9.2in howitzer, and this also became a British standard equipment until the 1940s. Finally, under the urging of Admiral Bacon retired from the Royal Navy to become its director. Coventry produced a 15in howitzer of which 8 were built and operated in France during the First World War by the Royal Marines. The company also manufactured large numbers of guns designed by other companies during the First World War and also made munitions and aircraft guns. Coventry folded during the armaments slump of the 1920s.

**CRADLE** The part of the gun that supports the barrel and carries the recoil mechanism. In addition, it usually carries the trunnions and the elevating arc. The cradle may be a 'trough' cradle, in which the recoil system is carried inside a box-like unit with the gun mounted above; or it may be a 'ring' cradle, in which case the gun passes through it and the recoil system cylinders are distributed above and below the cradle.

**CRANE LINER** An American development during the Second World War, the Crane Liner was a hardened steel insert shaped internally to the contour of the forward end of a gun chamber and the first few inches of rifling. A worn barrel could be cut away internally to remove scoring and erosion, and the Crane Liner inserted to provide a new forcing cone and commencement of rifling. In this way it was possible to prolong the life of a gun or howitzer that was otherwise in good condition. The same idea had been used in British guns during the First World War when it was known simply as a 'short bore and chamber liner', but although the idea sounds attractive, it has serious drawbacks. One is that the liner develops a tendency to twist due to the resistance to the driving band of the rotating shell, and this can, in time, become serious enough to

cause premature detonation of the shell. Another drawback is that the metal has a tendency to 'creep' forward and build up a choke at the front end, and so be another possible cause of prematures.

**CREST CLEARANCE** An element in the controlling of fire by artillery in the field, to ensure that the shell, on its way to the target, does not strike any intervening crest en route, especially if the crest happens to be occupied by friendly troops. The angle of elevation necessary to clear all intervening crests had to be calculated by trigonometry, after which care was taken that no fire took place below this angle. The requirement is the same today but the calculation is performed by computer.

**CROSS-AXLE TRAVERSE** A method of traversing the gun on its carriage by forming the outer surface of the axle into a screw-thread. A nut, connected to a handwheel and to the traversing mass, engages in this screw-thread, and when revolved by the handwheel it causes the traversing mass to move across the axle. Cross-axle traverse is only suitable for light guns, as the wheels of heavy

guns tend to settle into the ground when firing, and the movement of the traversing mass, pivoting on the trail spade, tries to push and pull on the wheels during traversing. When the wheels are sunk into the ground, this resistance cannot be overcome by the gear ratio of the cross-axle mechanism.

**CROSS-LEVEL** An adjustment to the sight of a gun to compensate for any lack of level of the gun due to the ground upon which it is placed. If the gun trunnions are not level, the gun as it elevates will also move in azimuth, so upsetting the pointing. The cross-levelling gear on the sight will bring the sight upright independently of the level of the trunnions, and this will automatically compensate for the movement of the gun barrel. Cross-level correction is no longer applied by the gunlayer in modern equipments, but is automatically applied to the sight data by an electronic sensing device in the gun mounting.

**A typical course and speed sight mounted on a German light anti-aircraft gun.**

**Below: The ring cradle removed from a British 3.7in AA gun. Note the trunnion on the right side and the elevating arc beneath.**

**A trough cradle on a German 105mm howitzer.**

**CRUCIFORM CARRIAGE** A type of carriage found with anti-aircraft and a few anti-tank guns. The gun sits on a pedestal attached to a central platform which has four arms equispaced around it. Outriggers are generally attached to these arms, and can be unfolded to give a greater span to the arms and thus more stability to the gun in action. The design is intended to allow the gun to rotate through 360° and fire in any direction, the carriage arms giving stability and resistance to the recoil no matter where the gun points when it fires.

**CUT-OFF GEAR** An automatic adjustment in the control valves of a recoil buffer (qv) which shortens the recoil movement as the gun elevates. Cut-off gear is usually found with howitzers and fitted so that the barrel is allowed the maximum recoil when horizontal and progressively shorter recoil as it elevates to ensure that the breech does not strike the ground.

## CZECH ARTILLERY EQUIPMENT

| Equipment | Date | Calibre (mm) | Barrel length (cals) | Breech mech. | Elevation max (deg) | Traverse (deg) | Weight in action (kg) | Shell weight (kg) | Muzzle velocity (m/sec) | Range, max (metres) |
|---|---|---|---|---|---|---|---|---|---|---|
| **Anti-tank artillery** | | | | | | | | | | |
| 37mm vz34 | 1934 | 37 | 39 | | 30 | 50 | 275 | 0.9 | 675 | 5000 |
| 37mm vz37 | 1937 | 37 | 48 | | 26 | 50 | 370 | 0.9 | 750 | 5000 |
| 4cm vz30 | 1930 | 40 | 71 | HSB | | | | 1.05 | 950 | 9140 |
| 47mm vz38 | 1938 | 47 | 44 | | 26 | 50 | 570 | 1.7 | 775 | 5800 |
| 85mm vz52 | 1952 | 85 | 57 | HSB/SA | 38 | 60 | 2095 | 9.2 | 820 | 16200 |
| 6.6cm vz45 | Proto | 66 | 53.5 | VSB/SA | 20 | 65 | 1496 | 5 | 800 | |
| **Anti-aircraft artillery** | | | | | | | | | | |
| 4.7cm vz37 | 1938 | 47 | | VSB/SA | 85 | 360 | | 1.5 | 800 | 7100 ceiling |
| 7.5cm Kan PP Let vz32 | 1932 | 75 | 50 | VSB/SA | 85 | 360 | 2545 | 6.5 | 820 | 9750 ceiling |
| 7.6cm Kan PP Let vz28 | 1928 | 76.2 | 54 | | 85 | 360 | 5456 | 8 | 800 | 11400 ceiling |
| 8.3cm Kan PL vz22 | 1925 | 83.5 | 55 | HSB/SA | 85 | 360 | 8250 | 10 | 800 | 11000 ceiling |
| **Mountain artillery** | | | | | | | | | | |
| 10cm Mtn How vz16/19 | 1920 | 100 | 24 | HSB | 70 | 5 | 1350 | 15.8 | 395 | 9600 |

**The lower carriage of a 3.7in anti-aircraft gun, showing the construction of a cruciform mounting.**

**CZECHOSLOVAKIAN ARTILLERY** Czech artillery really means Skoda artillery, as the Skoda factory in Pilsen, supplier to the Austro-Hungarian and other armies, found itself part of Czechoslovakia when that country was created in 1919. The factory still exists and produces artillery and armoured vehicles for the Czech Army and other Warsaw Pact forces, but is now known as 'V.I. Lenin Works'.

When the Czech Army was formed it was originally outfitted from stocks of Austro-Hungarian weapons ceded to it, all of which were of Skoda make. Meanwhile Skoda was evaluating the lessons of World War One and began development of new designs, though because of shortages of finance the Czech Army merely had some of its older guns modified to give them better performance. In the mid 1930s money was made available

and the Army began receiving new weapons that represented the fruit of Skoda research and export sales experience, though even then the need to utilize existing stocks of ammunition meant that there was no radical change of calibre.

After the German takeover of Czechoslovakia in early 1939, Skoda was given German Army contracts for a variety of equipment, but still found time for development, producing experimental heavy mortars, rapid-firing anti-tank guns and field artillery. One of the most notable designs was the German 105mm leFH43 field gun, which was mounted on a cruciform carriage to permit all-round fire.

Since the integration of Czechoslovakia into the Warsaw Pact, most of the artillery has been of Soviet design, though one gun, the 85mm M52, is still made in the former Skoda factory.

### Czech Artillery Equipment (Cont)

| Equipment | Date | Calibre (mm) | Barrel Length (cals) | Breech mech. | Elevation max (deg) | Traverse (deg) | Weight in action (kg) | Shell weight (kg) | Muzzle velocity (m/sec) | Range max (metres) |
|---|---|---|---|---|---|---|---|---|---|---|
| **Field Artillery** | | | | | | | | | | |
| 7.5cm Field Gun vz19 | 1919 | 75 | 29 | HSB | 16 | 7 | 940 | 6.5 | 500 | 6000 |
| 7.5cm Field Gun vz35 | 1934 | 75 | 21 | HSB | 45 | 50 | 1040 | 6.3 | 480 | 10200 |
| 7.6cm Field Gun vz17 | 1919 | 76.5 | 30 | | 45 | 8 | 1386 | 8 | 500 | 9900 |
| 7.6cm Field Gun vz39 | 1939 | 76.5 | 30 | HSB | 45 | 50 | 1425 | 7.3 | 570 | 12000 |
| 8cm Field Gun vz17 | 1918 | 76.5 | 25 | HSB | 45 | 7 | 1335 | 8 | 500 | 10000 |
| 8cm Field Gun vz30 | 1934 | 76.5 | 40 | HSB | 80 | 8 | 1816 | 8 | 600 | 13500 |
| 85mm Field Gun vz52 | 1953 | 56 | 38 | VSB | 38 | 60 | 2095 | 9.3 | 805 | 16200 |
| 10cm Field Gun vz35 | 1936 | 105 | 42 | HSB | 42 | 50 | 4200 | 18 | 730 | 18400 |
| 10cm Field Gun vz53 | 1954 | 100 | 60 | HSB | | | 3400 | 15.7 | 900 | 21000 |
| 10cm Field How vz14/19 | 1923 | 100 | 24 | HSB | 50 | 6 | 1548 | 16 | 395 | 9800 |
| 10cm Field How vz30/34 | 1934 | 100 | 25 | | 70 | 8 | 1766 | 16 | 430 | 12200 |
| 10cm Field How H3 | 1939 | 100 | 30 | HSB | 70 | 50 | 1960 | 14.4 | 525 | 12200 |
| | | | | | | | | | | |
| **Heavy artillery** | | | | | | | | | | |
| 15cm How vz14/16 | 1919 | 149 | 14 | | 70 | 8 | 2930 | 42 | 350 | 8650 |
| 15cm Field How vz15 | 1919 | 150 | 20 | | 65 | | 3800 | | 500 | 11880 |
| 15cm How vz25 | 1925 | 149 | 18 | | 70 | 7 | 3800 | 42 | 450 | 11800 |
| 15cm How vz37 | 1937 | 149 | 24 | IS | 70 | 45 | 5200 | 42 | 580 | 15100 |
| 15cm Gun vz15/16 | 1919 | 150 | 40 | | 45 | 6 | 16480 | 54.4 | 700 | 20100 |
| 21cm How vz18 | 1919 | 210 | 10 | | 70 | 360 | 9042 | 135 | 380 | 10100 |

### CZECH SELF-PROPELLED ARTILLERY

| Equipment | Date | Calibre (mm) | Chassis | Elevation max (deg) | Traverse on mount (deg) | Weight in action (kg) | Range, max (metres) | Ammunition carried (rds) |
|---|---|---|---|---|---|---|---|---|
| DANA SP Howitzer | 1980 | 152 | 8-wheeled | 60 | 360 | 23000 | 20000 | |

# D

**DANGLIS** A Greek army officer who was associated with the Schneider company of France in the period 1900-14 and who produced a number of designs of mountain and field guns which were adopted by the Greek Army under the designation 'Schneider-Danglis'.

**DASHIELL BREECH** A continuous-motion (qv) breech mechanism invented by Lieutenant R.B. Dashiell, US Navy, c.1895, and used on 4, 5 and 6in US naval guns firing cased charges. A hand lever extended across the rear of the breech and had teeth on its shorter end; as the lever was pulled outwards, so these teeth pushed a rack bar across to unscrew the breech block. At the end of this movement a 'translating arm' withdrew the breechblock on to a tray, after which further movement of the handle swung the tray and block open to allow the gun to be loaded. An unusual feature of this design was that the screw revolved left-handed in order to lock.

**Above: The version of the Schneider-Danglis mountain gun of 1910 adopted by the Greek Army.**

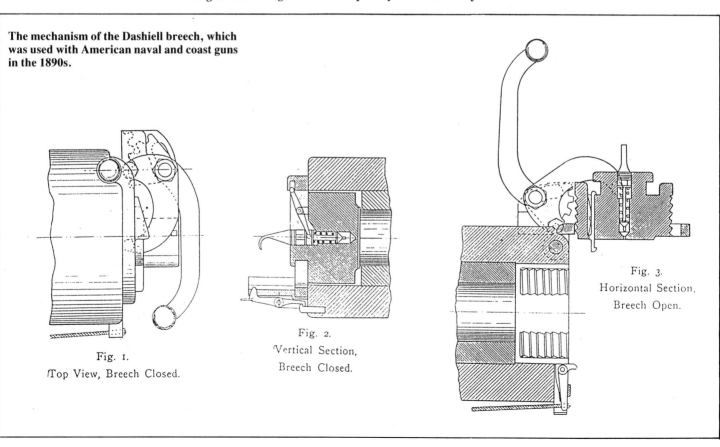

**The mechanism of the Dashiell breech, which was used with American naval and coast guns in the 1890s.**

Fig. 1.
Top View, Breech Closed.

Fig. 2.
Vertical Section,
Breech Closed.

Fig. 3.
Horizontal Section,
Breech Open.

**DATUM POINT** The British term for 'witness point' (qv), a point upon which a gun could be ranged to determine corrections due to weather conditions, etc, after which the deduced data could be applied to a target nearby.

**DEAD TIME** An expression used in anti-aircraft gunnery, signifying the time between the fuze being set and the gun being fired. In the early days an arbitrary dead time was used in calculations, but the actual dead time was variable and depended on how long it took a man to set the fuze by hand and then load the gun, which could thwart an accurate prediction of the shell's position in the sky when it burst. The development of mechanical fuze setter-loader devices fixed the dead time which improved the accuracy of prediction and hence the effectiveness of fire. Dead time was made obsolete by the introduction of radio proximity fuzes, which required no setting and burst when they sensed the proximity of the target.

**DE BANGE** A French inventor, Colonel de Bange perfected a system of obturation for bag-charge guns which has been in use since the 1880s. It relies upon a resilient pad held between the face of the breech block and a 'mushroom head'. On firing, the mushroom head is driven back by the pressure of the explosion and the pad is squeezed sideways against a prepared surface on the inside of the chamber wall. A pressure differential ensures that the sideways pressure is always in excess of the chamber pressure and so guarantees an efficient seal.

**DEPORT** An artillery officer of the French Army, Deport was one of the design team that developed the 75mm M1897 gun. After reaching the rank of Colonel he retired from the army in the late 1890s and became associated with the Compagnie des Forges de Chatillon of Montluçon in France. The company developed a number of field and mountain guns under his guidance which were sold to various countries, as well as being adopted in small numbers by the French Army. Deport's most significant contribution to artillery design was the split-trail carriage (qv), which he designed for a 75mm gun in 1912 and which has subsequently been adopted all over the world. He also invented the dual recoil system (qv), which he applied to the same gun, and the differential recoil system (below), applying it first to a 60mm mountain gun adopted by the French Army in 1907. This met with less success, though the principle has been adopted occasionally since then.

**DIFFERENTIAL RECOIL** A system of recoil control intended to relieve stress on the gun and reduce the weight of the recoil system and carriage, invented by Colonel Deport of France for a 60mm mountain gun in 1907. The principle employed is to haul the gun back against its recuperator spring and lock it there. The gun is then loaded. On pulling the firing lever the gun is released to run forward in its cradle. The cartridge is fired a fraction of a second

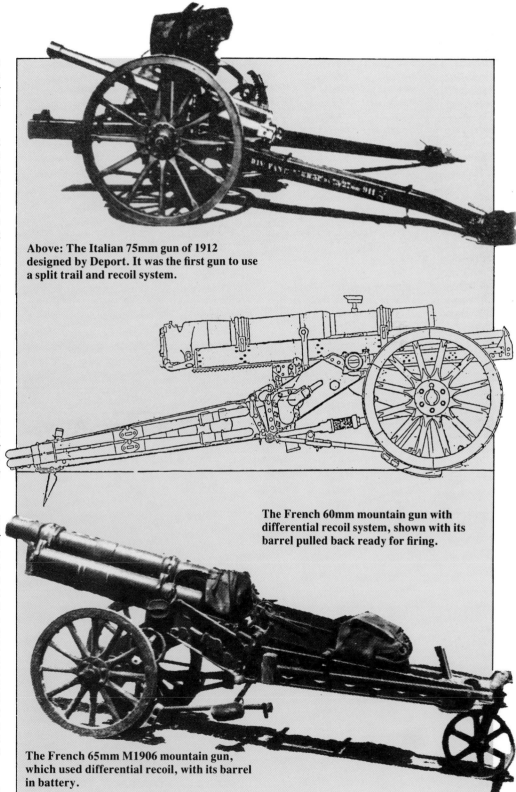

Above: The Italian 75mm gun of 1912 designed by Deport. It was the first gun to use a split trail and recoil system.

The French 60mm mountain gun with differential recoil system, shown with its barrel pulled back ready for firing.

The French 65mm M1906 mountain gun, which used differential recoil, with its barrel in battery.

before the gun reaches the end of its run. Therefore the recoil has first to arrest the moving mass, then reverse it and make it recoil again, and the barrel is again caught and held at the fully-recoiled position ready for re-loading.

The drawback to this system is that if the gun misfires it is liable to tip forward onto its muzzle,

and if the cartridge then fires the results are interesting to say the least. The idea was revived during World War Two by the Germans for a 55mm anti-aircraft gun, and was later re-revived by the US Army in the 1970s, at which time it was known as 'soft recoil', though this project was later abandoned.

**DIRECT FIRE** Fire from a gun which can see the target and take a direct aim against it; as, for example, in anti-tank and tank guns. Until the latter part of the 19th century, all artillery fire was direct, but the development of high-powered infantry rifles led to artillery having to conceal itself on the battlefield, and from this came the perfection of indirect fire (qv).

**DIRECTOR** 1, Optical instrument capable of measuring angles in azimuth and elevation, used for simple surveying and orientation of guns prior to firing. (British and French terminology; the same instrument is called an 'Aiming Circle' in the USA.) 2, An anti-aircraft fire control computer (American); see 'Predictor'.

**A modern British Army director, used for aligning guns and other tasks that demand the measurement of vertical and horizonal angles.**

**DRIFT** The sideways movement of a shell as it passes through the air, caused by its rotation due to the rifling. Drift is caused by a combination of three forces: equilibrium yaw, Magnus effect and Coriolis effect. Equilibrium yaw refers to the gyroscopic tendency of a spinning shell to maintain the same attitude in space while gravity is causing its flight path to curve; this means that the nose will not be exactly on the trajectory but lying slightly to the right (in the case of right-hand spin), and that air pressure will push the shell further to the right. Magnus effect is caused by the spinning shell carrying air round with it; this meets the oncoming air flow, and the net effect is, again, to push the shell off course. Coriolis effect is an error due to the shell's motion being considered as being fixed in space when, in fact, it refers to the rotating earth. Drift is always in the direction of the twist of rifling, and for a given gun and shell combination is constant so that it can be compensated for in the design of the sight.

**DISAPPEARING CARRIAGE** A type of gun carriage used for artillery of position, which it is desired to conceal between shots. It is principally used with coast defence guns, but occasionally with mobile siege and fortress guns.

Numerous attempts were made in the 18th century to develop a carriage that would conceal the gun but allow it to rise above cover for firing, but the first practical model was due to Captain Moncrieff of the Edinburgh Militia Artillery, which was patented by him in 1863 and later adopted by the British and other armies. The gun (muzzle-loader) was mounted at the top of two curved arms which carried a large counter-weight at their foot. The gun recoiled, on firing and caused the arms to roll back on their curved surface, so lifting the counter-weight. The gun was held down by a friction brake while it was loaded; then the brake was released and the counter-weight rolled the arms back, lifting the gun above the parapet of the pit in which it was installed.

The Moncrieff carriage was a success with light guns, but was not suitable for calibres greater than about 7 inches. Moncrieff then went to the Elswick Ordnance Company and, with its help, developed a design in which the gun arms were pivoted at their foot, on a revolving structure, and resistance to recoil was provided by an hydraulic ram that was driven in as the gun recoiled. Mountings of this patterns were adopted for guns of up to 13.5in calibre in British service, and similar mountings were developed in other countries. The US Coast Artillery developed the Buffington-Crozier (qv), which carried guns up to 16in calibre.

The object of the disappearing carriage was to conceal the gun so that enemy ships could not fire at it. However, trials in Britain in the 1890s showed that the fire from ships' guns was extremely unlikely to hit any gun on shore, because the target was so small, and with the development of barbette mountings with recoil control, which were much cheaper, the disappearing carriage was rendered obsolete. Disappearing carriages were withdrawn from British service shortly before the First World War, though a few lingered into the 1930s until they were replaced by more modern carriages. The US Army retained them in use throughout World War Two, largely because the guns were still effective and money for their replacement had not been forthcoming in the years between the wars.

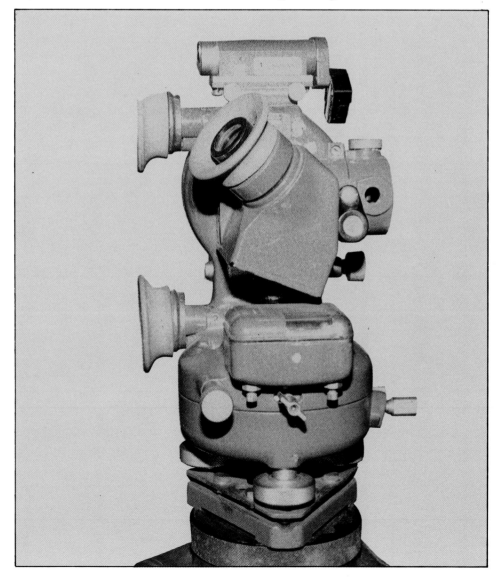

**Two American 10in disappearing guns, one 'up' and one 'down' are preserved at Fort Casey, Washington State.**

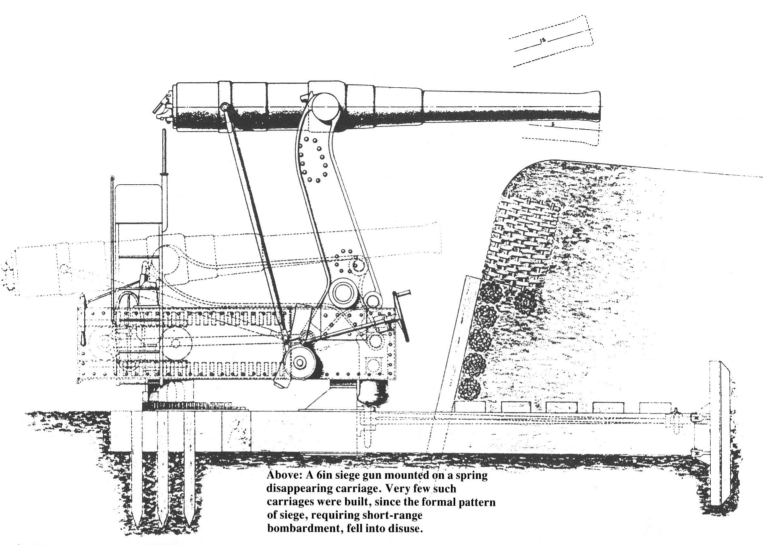

Above: A 6in siege gun mounted on a spring
disappearing carriage. Very few such
carriages were built, since the formal pattern
of siege, requiring short-range
bombardment, fell into disuse.

**A German 17cm gun showing the practical application of a dual recoil system in which the top carriage and the barrel recoil.**

**DUMARESQ** A fire control instrument devised originally for the Royal Navy, then adopted by British and other coast defence artillery organizations. In essence it was a simple computer which would produce rates of change in both range and bearing when fed with the bearing, speed and range of the enemy, together with elapsed time. It was possible from this data to deduce the future position of the target at any chosen time and to fire guns so that the shell arrived at the projected position at the same time as the target.

**DUMMY LOADER** A training device for teaching recruits how to load the gun, it consists of the breech of the gun in question attached to a length of tube that represents the barrel, supported on a

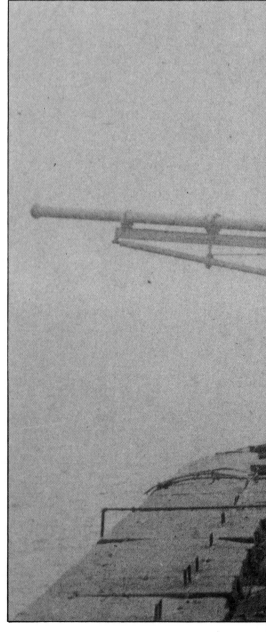

**DISTRIBUTION** The practice of applying corrections to the azimuth of the guns of a battery to enable them to spread their lines of fire and so distribute their shells along a frontage greater than that of the battery.

**DREYER TABLE** A fire control instrument devised for the Royal Navy and later taken into use by coast defence artillery. It consisted of a table with various mechanical computing devices coupled to a pen moving across a chart or map. When fed with the target's initial range and bearing, and with rates of change in these from the Dumaresq (qv), the pen plotted the present and future positions of the target, from which firing data could be deduced. Similar devices, under other names, were used by coast artillery of several nations. They were generally replaced by electro-mechanical computers during the 1940s.

**DRIGGS-SEABURY** An American armaments manufacturing company active in the period 1890-1940; known as the Driggs Ordnance and Engineering Company in its latter years. It built a number of small quick-firing coast defence guns for the US Army in the early years of the century, but thereafter appears to have dealt only in small export orders for armament while expanding its business in the general engineering line.

**DROOP** The curvature of a gun barrel in the vertical plane, due to two components: gravity deflection and bend from errors in manufacture. Droop was particularly pronounced in wire-wound guns (qv), which had lesser longitudinal strength

than built-up guns. Generally the same for all guns of the same model, droop is measured during the initial testing of the gun and is thereafter compensated for in the firing tables.

A variation of droop is the amount of bend due to the gun heating up from firing or from the heating effect of the sun which, of course, acts on one side of the gun and thus pulls it into a bend. Tank guns are today wrapped in thermal blankets to counteract these effects and to hold them at a steady temperature.

**DUAL RECOIL SYSTEM** A system of recoil control used with some heavy guns. The gun is carried in the usual type of cradle, attached to the top carriage (qv) and permitted to recoil in the normal manner. But, in addition, the top carriage itself can recoil on the lower carriage, being controlled by the normal sort of hydro-pneumatic recoil mechanism. Thus the recoil force is divided between the two systems, the cradle-mounted system absorbing part and the top carriage system absorbing the remainder. It was invented by Captain Deport (qv) for a 75mm gun in 1912, but thereafter saw little application until the late 1930s when it was used by Germany in the 17cm Kanone 18 and the 21cm Howitzer, both of which were designed to interchange on the same carriage as 'partner pieces'. An interesting modification was seen on the German 21cm K12 railway gun, where the gun recoiled in its cradle and the entire mounting recoiled across its rail trucks. The use of dual recoil reduces the blow to the gun platform, and it makes the lower carriage extremely steady during firing.

frame at the correct height above the ground. The gun detachment form around it in the correct positions and learn the various steps involved in loading, having to operate the breech mechanism, load the shell and ram it, load the cartridge, load the firing lock if necessary, and finally close the breech and operate the firing gear. All the ammunition used is, of course, dummy. After ramming, the shell falls out of the end of the tube and is returned to the ammunition pile either by some form of mechanical runway or by a spare man or men carrying it. In the case of bag-charge guns the dummy cartridge is similarly pushed out of the tube by the ramming of the next shell; with cased-charge guns the opening of the breech extracts and ejects the cartridge in the correct manner.

Dummy loaders are generally credited as an invention of Captain Percy Scott, the Royal Navy's famous gunnery expert in the 1890s, though there seems a strong possibility that this merely refers to the official manufacture and issue of the device; various evidence suggests that army batteries had already developed similar devices as local expedients on an unofficial basis.

**DYNAMITE GUN** A gun using compressed air as the propellant, developed in the 1880-90 period in the United States. Originally devised by a Mr. Mefford, the idea was taken over by Lieutenant Zalinski of the Coast Artillery Corps and becme known as the Zalinski Dynamite Gun. The 'dynamite' part of the name came from the shell's filling, which was dynamite, and it was this that led to the use of air as a propelling agency. A dynamite-filled shell fired from a conventional gun would have detonated in the bore due to the shock of discharge, but the air propulsion gave the shell a more gentle push. The gun was intended as a coast defence weapon, to pitch the shells into a harbour to reinforce an existing minefield. A few were installed in the USA, but before long the development of explosives capable of being fired by conventional artillery rendered the weapon obsolete.

**The Zalinski Dynamite Gun installed as part of the harbour defences of New York in the early 1890s.**

# E

**EFCs** An abbreviation for 'Effective Full Charges', which is a measure of the probable life of a gun by referring to the number of full charge cartridges it will be able to fire before being worn out. Most of a gun's firing is done with reduced charges, so before the actual cartridges fired can be converted into EFCs it becomes necessary to decide how many of these represent a full charge. The life in EFCs depends entirely upon the power and velocity of the gun; a field howitzer, lightly stressed, may well have a life of 15,000 or more, whereas a high velocity tank gun may have a life of only 500 EFC.

**EHRHARDT** Dr. Heinrich (1840-1928) was the gun designer and chief ordnance engineer of the Rheinische Metallwaaren- und Maschinenfabrik of Sömmerda and later Düsseldorf, which later became better-known as Rheinmetall. He developed a number of field and mountain guns with on-carriage recoil in the 1895-1914 period which sold widely throughout the world, and was responsible for the 15pr QF gun which was purchased by the British Army in 1901 for service in South Africa.

**EIGHTY-EIGHT** The popular name for the highly effective multi-purpose German anti-aircraft, anti-tank and tank gun during the Second World War. It was originally designed in the late 1920s by Krupp technicians working with Bofors of Sweden; they returned to Germany in 1931 and a prototype gun was built and demonstrated. It was immediately approved and went into production, being introduced into service in 1933 as the 8.8cm FlaK18. It was a straightforward air defence gun on a cruciform platform with four wheels that were removed before firing. The breech was a semi-automatic sliding block type, and sights were provided for on-carriage use or for the receipt of data from a central predictor. It fired a 9.5kg (21lb) shell at 820m/sec (2,690ft/sec) to a maximum ceiling of 9,900 metres (32,480ft) and an effective ceiling of 8,000 metres (26,247ft) and could achieve a rate of 15 rounds per minute with a skilled detachment. As a ground gun it had a maximum range of 14.8 kilometres (9.2 miles).

After some experience had been gained in the hands of troops, changes were made to simplify manufacture and improve operation of the mounting. The gun was changed to a three-piece liner construction so that the three parts – chamber, centre and muzzle – could be of different types of steel, allowing the use of cheaper steel where the wear was less and permitting the replacement of only the part of the barrel that was worn or damaged. The ballistic performance remained the same; new sights were adopted, and this new equipment became the 8.8cm FlaK36, being introduced in 1937.

Numbers of both types were sent to Spain during the Civil War and their capabilities as anti-tank and field artillery weapons were then discovered. As a result they were provided with direct-fire sights and anti-tank projectiles, though the 88's full potential as an anti-tank weapon did not make itself apparent until the battles against the British Army in the Libyan Desert in 1941-2.

In view of the increasing performance of aircraft a specification for a better gun was issued by the Luftwaffe in 1939. This resulted in the Rheinmetall-Borsig design known as the 8.8cm FlaK41, which entered service in 1943 after some initial teething troubles. This differed considerably from the FlaK18, principally in having the gun mounted on a turntable carried on a four-legged platform. The gun had its trunnions at the rear of the cradle, well behind the breech, allowing it to elevate to 90°, and it fired the same 9.4kg (21lb) shell to a maximum ceiling of 15,000 metres (49,213 feet) and an effective ceiling of 11,000 metres (36,089 feet). Its ground range was 19,700 metres (21,544 yards). A few were used in Tunisia, but most of these weapons were kept for the defence of Germany.

Krupp had also received a contract to develop an anti-aircraft gun; without getting too involved in details, it can simply be said that as a result they began to attempt a 'family' of 88mm weapons for air defence, anti-tank and tank gun roles. The air defence plan was dropped when the Rheinmetall design was adopted, and Krupp thereafter concentrated on the anti-tank and tank patterns. The anti-tank gun became the Pak (Panzer Abwehr Kanone) 43 and entered service in 1943.

**Above: The German 88mm FlaK 18 gun prepared for travelling.**

**Left: Ehrhardt was an innovative designer. This is his 14pr anti-balloon motor gun of 1909.**

**Left: The 88mm FlaK 18 gun in firing position.**

The gun was of two-piece construction with a vertically-sliding semi-automatic breech, and was mounted on a cruciform carriage to allow 360° traverse and 40° elevation. With the wheels removed it sat very low to the ground and could be easily concealed. It could also be fired off its wheels in an emergency. The gun fired a 10.4kg (23lb) piercing shell at 1,000m/sec (3,281ft/sec) and could defeat 16cm (6½in) of solid armour at 2,000 metres (2,187 yards) range, a formidable performance for its day. With tungsten-cored shot fired at 1,130m/sec (3,707ft/sec) it could defeat 184mm (7in) of armour at 2,000m (2,187 yards) range. It could also fire high a high explosive shell to a range

of 17.5 kilometres (11 miles), giving it a useful support capability.

The appearance of the Soviet T-34 and KV tanks, which appeared to be impervious to the German 5cm anti-tank guns, led to increased demands for the 88mm Pak43, and as barrel production was faster than carriage production, an emergency version known as the 8.8cm PaK43/41 was adopted, using a conventional two-wheel split-trail carriage developed from existing components of other weapons. It was somewhat cumbersome, and was christened 'Scheunetor' (Barn-door) by the German troops in Russia who had to push it through the mud, but it was a devastating weapon;

there are authenticated records of one gun which knocked out six T 34 tanks at a range of 3.5 kilometres (2 miles), and another which, at 600 metres (656 yards), blew the engine block 5 metres (5½ yards) and the turret 15 metres (16 yards) away from the hull of the tank.

There were two 88mm tank guns, the 8.8cm KwK (Kampfwagen Kanone) 36 which, as the title suggests, was the FlaK36 gun, and the 8.8cm KwK43, which was the PaK43 gun; both were mounted in Tiger tanks. Numbers of Pak 43 barrels were also mounted into self-propelled chassis to form tank destroyers.

131

**ELECTRIC GUNS** Devices for launching projectiles by means of electric current. The basic principle is of a solenoid (ie, a coil of wire), which has a current induced and gives rise to a magnetic field that ejects a soft iron slug placed in the centre of the core. Several inventors in the 1880-1900 period took out patents for guns which used this principle, but none was successful because sufficient velocity could not be obtained to launch a worthwhile projectile. In 1917 a French inventor, Fauchon-Villeplé, designed a 'gun' that consisted of two parallel rails to which current was applied. The projectile was a dart with wings that rested on the rails and was launched at high speed by the magnetic field. Fauchon-Villeplé was given a contract to develop a workable gun. He had not achieved success by the time of the Armistice and the contract was terminated, but his idea was worked upon by others in later years and eventually produced the 'linear motor'.

In 1940 Herr Muck, a German engineer, proposed a battery of enormous electric guns which worked on the solenoid principle. These were to be buried in a hillside close to the Belgian border and adjacent to coalfields (necessary to feed the power stations that would have to be built to supply the electric current) and were to be used to bombard London. The idea was attractive to Hitler, who loved grandiose schemes, but when subjected to close scientific analysis it was found to be impracticable and was dropped. In 1944 the German 'Gesellschaft für Gerätebau', under Flight-Engineer Hänsler, put forward a proposal for a multi-barrelled 4cm calibre anti-aircraft gun using a modified linear motor, and promised a weapon that would fire a 500g (1.1lb) shell at a muzzle velocity of 6,500ft/sec (1,981m/sec) at a rate of 6,000 rounds per minute and requiring 3,900 kilowatts of power. The Luftwaffe were sufficiently impressed to give Hänsler a development contract, but the war ended before he got very far. In postwar years the idea was closely examined by British scientists who came to the conclusion that each gun would have required a power station of a size sufficient to supply a medium-sized city, and the idea was again abandoned.

The idea returns from time to time, and as electrical engineering and science improve, gets

**Below: An early type of elevating screw on an 18th century field cannon.**

**Right: Elefant, also called Ferdinand, the German tank destroyer.**

closer to realization with each appearance. At present the 'electric gun' appears to be taking the form of a laser projector for use in space.

**ELEFANT** A German tank destroyer – ie, a self-propelled anti-tank gun – formally known as 'Sturmgeschütz m/8.8cm PaK43/2, Sd Kfz 184', less formally as 'Ferdinand' after Dr. Ferdinand Porsche, the designer.

Porsche had designed a version of the Tiger tank which, due to technical problems with the petrol-electric drive, was not selected for production. In late 1942, it was decided to develop a heavy tank destroyer mounting an 88mm gun, and the Tiger (P) (as Porsche's design was known) was taken as the basis. The resulting vehicle had a fixed superstructure mounting the 88mm gun facing forward, was protected by 200mm (9in) of armour and weighed 64 tonnes. Ninety were built in time for use in the Kursk offensive in the summer of 1943, and later were used in small numbers on the Italian front. They were formidable weapons but, like all non-turreted armoured vehicles, were vulnerable to attack from the flanks and rear.

**ELEVATING GEAR** The apparatus for elevating the barrel of the gun. Elevating gear is generally a handwheel which turns a worm gear engaging with a curved toothed rack attached to the elevating portion. On fixed fortress guns and on self-propelled guns and tanks the gear can be power-actuated, though hand operation is always provided for emergency use. A method occasionally used is to force the gun barrel up and down by means of an hydraulic ram.

**ELSWICK ORDNANCE COMPANY** The gun-making company set up by Sir William Armstrong (qv) to manufacture the Armstrong gun under contract to the British Army and later for export in the 1860s. It remained in business and moved on to manufacture rifled muzzle-loading guns and then breech-loading guns. The name, derived from the locality close to Newcastle-upon-Tyne, vanished when Armstrong amalgamated with Vickers in 1927. Note that the correct local pronunciation is 'Elzik'.

**EMPLACEMENT** A prepared site from which a gun is fired, and generally taken to refer to a masonry or concrete structure in a permanent location such as a fort or coast defence battery or for an anti-aircraft gun. It consists of the gun floor, on which the gun mounting rests, a parapet that protects the gun and its detachment from enemy observation and fire to some extent, and 'ready-use magazines' to hold ammunition for the gun. Additional ammunition is supplied from a magazine by lifts or other means.

**EPEE** An expression derived from the French term for a light curved-blade and referring to a curved railway track from which a railway gun is fired. Railway guns in the very large calibres (above 8 inches) have little traverse available on the mounting, and in order to point the gun at different targets it is pushed or pulled along the curved track until the barrel points in the desired direction.

Fine pointing is then performed by the on-mounting traverse.

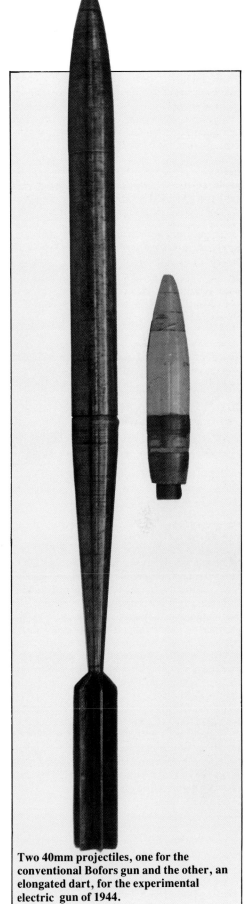

**Two 40mm projectiles, one for the conventional Bofors gun and the other, an elongated dart, for the experimental electric gun of 1944.**

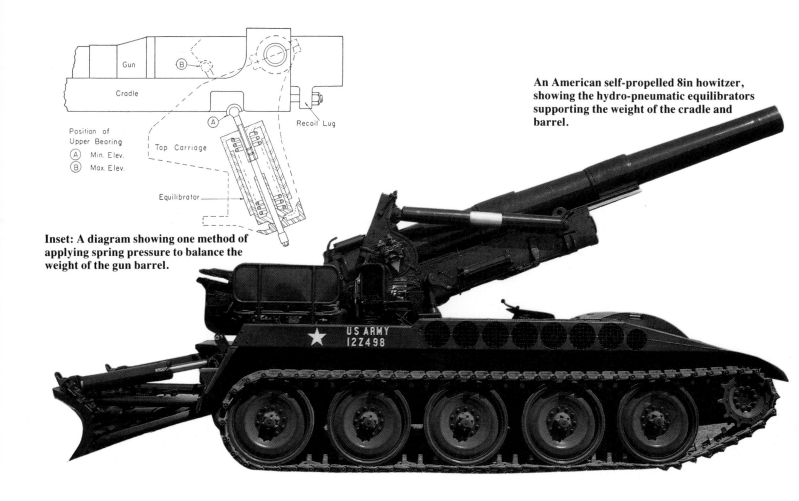

Position of
Upper Bearing
Ⓐ Min. Elev.
Ⓑ Max. Elev.

**Inset: A diagram showing one method of applying spring pressure to balance the weight of the gun barrel.**

**An American self-propelled 8in howitzer, showing the hydro-pneumatic equilibrators supporting the weight of the cradle and barrel.**

**EQUILIBRATOR** or Balancing Gear. This is the apparatus that is placed between the elevating mass of the gun and the fixed portion of the carriage in order to support the muzzle preponderance and so balance it that the gun can be elevated or depressed with relatively little effort on a handwheel. The equilibrator generally consists of a spring or hydro-pneumatic cylinder that pushes or pulls on the cradle. If the gun trunnions were at the centre of balance, equilibration would be unnecessary, but modern guns require the trunnions to be at the rear so that the breech does not strike the ground when the gun is elevated.

**EXTRACTOR** A claw-like metal component in the entrance to the gun chamber. It rests against the rim of the cartridge case and, on opening the breech, pulls the case loose from the chamber and throws it clear. The extractor is usually operated by connection with the breech block, though there have been cases when it has been geared to the breech operating lever. The extractors also act as a lock for the breechblock in semi-automatic breech mechanisms, which they keep open against a spring until a fresh cartridge is loaded.

**The open breech of a mountain howitzer, showing the extractor claw lying at the mouth of the chamber.**

# F

**FH70** A 155mm auxiliary-propelled howitzer currently used by Britain, West Germany, Italy, Japan and Saudi Arabia. Development began as a joint US/UK/FRG programme in the 1960s, the abbreviation 'FH70' meaning 'Field Howitzer of the 1970's'. Differences of opinion arose, the USA left to develop its own design (the M198), Britain and West Germany agreed to go ahead in 1968, and in 1970 the Italians joined the programme. The first international trial battery of six guns was formed in 1975 and the first production weapons went into service in 1976. All British, German and Italian requirements were filled by 1982, and since that time production has continued on orders for Japan and Saudi Arabia. Britain is responsible for the carriage, traversing gear, high explosive shell and cartridge; Germany for the howitzer, loading system, auxiliary propulstion, suspension, sights and illuminating ammunition; and Italy for the cradle, recoil system, elevating mechanism and some ammunition items.

The FH70 uses a split-trail carriage with two main wheels that can be hydraulically driven by the auxiliary propulsion unit (APU). There are two smaller wheels at the trail ends, which steer the weapon when under its own power; these can be retracted when the gun is being towed by a tractor. The howitzer consists of a 39-calibre barrel with muzzle brake, fitted with a rising-block semi-automatic breech. The gun rides on a cradle in which the recoil system is fitted, and balancing gear supports the muzzle weight. The gun is swung through 180° when moving and clamped above the trails. The APU uses a Volkswagen air-cooled engine driving hydraulic pumps to provide power for movement, opening and closing the trail legs, loading, and raising and lowering the main and auxiliary wheels. The ammunition is separate loading and uses a standard 43.5kg (96lb) high explosive shell and combustible-cased bag-charge that is divided into eight zones. This shell gives a range of 24,700 metres (27,012 yards) at maximum charge. There is also an extended-range projectile which uses base bleed to increase the maximum range to 31,500 metres (34,448 yards). Other types of shell available include illuminating and smoke, and all existing types of NATO-standard 155mm ammunition can also be fired, though with less range available. The FH70 weighs 9,300kg (9 tons), is served by a detachment of seven or eight men, and can be towed by a 7-10 tonne 6 × 6 truck at speeds up to 100km/hr (62mph). When propelled by its APU it can reach a speed of 16km/hr (10mph).

**Firing the 155mm FH70 howitzer during trials in Sardinia.**

**FIELD ARTILLERY** A term that serves to classify those types of artillery used with a mobile field army and, more specifically, defines the weapons that are organic to the division. It generally refers to guns and howitzers of 75-105mm calibre, capable of being manned by about six men and moved by towing with a light truck or, in earlier days, by a team of six horses.

**FILLOUX** A French artillery officer and gun designer, Lieutenant-Colonel Filloux worked at the Puteaux Arsenal and designed two important guns in 1916-17. The first was the 'Canon de 155 long GPF', which became the standard French medium gun and was also adopted by the US Army in 1917; the second was the 'Canon de 194 GPF', which was a 194mm barrel mounted on the 155mm GPF carriage. GPF stands for 'Grand Puissance, Filloux'.

**FIRE FOR EFFECT** The fire of guns intended to have tactical effect against the enemy, as opposed to 'adjustment' or 'ranging' which is entirely for the purpose of determining the correct range and azimuth to permit fire for effect to take place.

**FIRING LANYARD** A cord attached to the firing lever of a gun. The purpose of the lanyard is to ensure that the firer does not have to get close to the gun to fire it, and thus does not run the risk of being hit by the gun as it recoils. In smaller equipments a firing lever is arranged so that the gun can be fired by the gunlayer, who is seated on the carriage.

**FIRING MECHANISM** The mechanical arrangement that actually fires the gun. In the case of guns using cartridge cases with a primer in the base, the firing mechanism is inside the breechblock and consists of a spring-driven striker that is released by a firing lever or lanyard. In tank and some field guns, the primer is fired electrically, using a spring-loaded contact button in the breechblock fed with electricity from batteries or a magneto. In bag charge guns, the firing mechanism is known as the 'firing lock' and is attached to the rear of the breech screw. It is loaded with a 'tube' or primer, which is similar to a blank cartridge for a rifle, and is fired either by a firing pin or electrically so that the flame from the primer passes down the vent and into the chamber to ignite the cartridge bag. Percussion firing mechanisms (ie, those using

a striker) need to have the striker withdrawn as soon as the breech begins to open so that it is not snapped off by sliding across the cartridge case, and must also have a means of re-cocking during the breech opening movement.

An unusual system is employed by the French 155mm GCT self-propelled howtizer; this has an induction coil built into the sliding breechblock, and the bag cartridge has a smaller induction coil connected to an electrical igniter inside the propellant. When the charge is loaded, the breech closes so that the breech coil is aligned with the coil in the bag charge. To fire, an electric current is made to flow in the breech coil; this induces a current to flow in the coil inside the cartridge, and this is sufficient to fire the electric igniter. The advantage of this system is that there is no vent in the breechblock to erode and, since the howitzer is loaded mechanically by remote control, there is no need to design a complex magazine for reloading primers into a firing lock.

**American troops practice firing 105mm howitzers in 1942.**

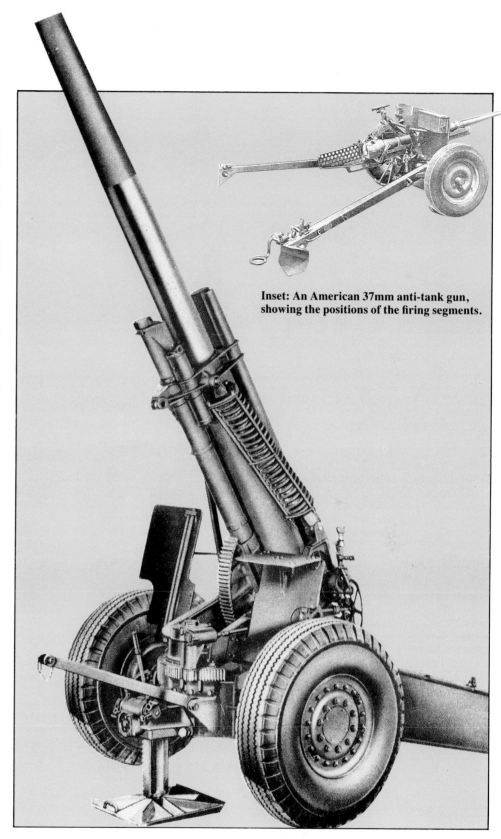

**A typical firing jack, lifting the wheels of an American 4.5in gun.**

**FIRING PEDESTAL/JACK** A jack or support lowered from beneath the axle of a field gun or howitzer in order to support the weight of the gun during firing and relieve the wheels and suspension of strain. Also, in the case of a split-trail gun, to give three-point suspension, so allowing the gun to take up a stable position on uneven ground.

**FIRING SEGMENTS** Arcuate metal supports pivoted to the axle of a gun and which, when allowed to swing down, press on the ground and relieve the gun wheels and suspension of strain during firing. Similar in effect to a firing pedestal (above).

**FIRING TABLES** Tables of figures that tabulate the range to be reached for any elevation of the gun when it is firing a specified charge and shell. The tables will also indicate the time of flight of the shell to the particular range, the gun's zone (qv) at that range, the change in range or lateral distance due to a specified change in azimuth or elevation at that range, fuze timing for the standard time fuze at that range, and corrections to be applied for wind, temperature and other effects. The information in the firing table is calculated from trial firings of the gun during its development, and a separate table has to be prepared for every possible gun, cartridge and shell combination. Today, most of the information in the firing table is carried in the artillery computer, but the table is necessary as a fall-back when the computer fails or is not available.

**FISA PROTECTORS** An American device to reduce wear in high velocity guns. The Fisa Protector was a hard steel sleeve that fitted round the mouth of the cartridge and the base of the shell. The chamber was relieved at its front end to accommodate the additional thickness. The protector took the force of the exploding gases and flame on firing and thus prevented erosion of the front of the chamber and commencement of rifling. The protector was then ejected with the empty cartridge case. The idea was developed in 1943-5 but was never entirely successful and was abandoned when the Second World War ended.

**FLETCHER BREECH** A type of continuous-motion (qv) breech mechanism for cased charge guns adopted by the US Army for the 3in field gun M1902 and for 4, 5 and 6in naval guns. It used a breechblock with four threaded segments and four plain, and also had a rack with teeth inclined at 45° to the axis of the block. A hand lever extended across the rear of the breech when closed; on pulling this lever outwards, the inner end, which had teeth, engaged in the rack and revolved the screw to unlock it. Once the screw was unlocked, further pulling on the lever pivoted the carrier which held the breechblock and swung it open.

**Inset: An American 37mm anti-tank gun, showing the positions of the firing segments.**

**FLICK RAMMER** A type of rammer (qv) used in modern self-propelled equipments. Instead of making a long, slow stroke to drive the shell through the chamber and engage it with the rifling, it gives a very short but intense impulse which gives the shell sufficient momentum for it to coast through the chamber and into the rifling. The rammer head folds up immediately after giving the impulse – to be out of the path of recoil – and unfolds, ready for the next round, after the breech has been opened. The object is to provide a rammer with very fast action so that the gun can load and fire a short series of rounds in a 'burst' of fire at the commencement of an engagement.

**Above: A Krupp 88mm anti-aircraft gun built for Argentina in 1938, showing the 'follow-the-pointer' dials on the side.**

**Right: French gunners cleaning 75mm Mle 1897 guns. Re-equipped with pneumatic tyres, the gun itself was essentially unchanged from 1897 to 1950.**

**FOLLOW-THE POINTER** A system of gun-laying adopted with artillery of position, which is supplied with constantly-changing data; eg anti-aircraft and coast artillery guns firing at moving targets. There are two gunlayers on the gun mounting, one controlling elevation and one controlling azimuth. Each watches a dial, which has a pointer, usually red, that moves according to signals transmitted from some central fire direction post. There is a second pointer, usually black, which is controlled by the movement of the gun, as controlled by the gunlayer. All the gunlayer has to do, therefore, is to keep moving his gun so that the black pointer matches the red; in practice, so that the black pointer covers and 'blacks out' the red pointer. So long as this is done, the gun is correctly laid according to the latest information and firing can take place at any time. A similar dual-pointer dial was attached to the fuze-setting machine, where fitted.

**FORCING CONE** An American term describing the conical shape of the front end of the gun chamber and the gradual rise of the lands above the grooves of the rifling. The forcing cone centres the projectile in the barrel at the start of its movement, and also grips the driving band of the shell when rammed so that it does not slip back into the chamber as the gun is elevated prior to firing.

**FRENCH 75** A popular name for the 'Canon de 75 Modèle 1897'. Developed by a design team led by Colonel Rimailho at Bourges Arsenal, and with a novel hydro-pneumatic recoil system developed at the Puteaux Arsenal, the French 75 was the pioneer of modern field artillery. This design brought together the features that were to be the characteristics of quick-firing field guns for many years afterwards: a recoil system that permitted the gun to remain steady on firing; a shield, behind which the men could be protected from small-arms fire; a rapid-acting breech mechanism; a fixed round of ammunition in which shell and cartridge were in a single unit that could be loaded in one movement; and the independent line of sight (qv) system of gunlaying. The M1897 fired a shrapnel shell weighing 7.24kg (16lb) to a range of 6.9 kilometres (4.3 miles). It was improved in later years, firstly by improved ammunition, then by the adoption of pneumatic tyres and finally by replacing the original pole-trail carriage with a split-trail pattern, all of which allowed the range to be increased to 11.1 kilometres (7 miles). The gun was adopted by the US Army in 1917 and by several other armies after 1918. It remained in use until 1943 with France and the USA and until the 1950s in some smaller armies.

The 25mm Hotchkiss anti-tank gun Mle 1934, here manned by British troops in 1940.

**FRENCH ARTILLERY** Although gunmaking had been carried out in government arsenals since the Middle Ages, in the mid 19th century private manufacturers began to dominate French procurement, largely because the military establishment had fallen badly into neglect. The two large companies of Schneider (qv) and St Chamond (qv) were responsible for much of the design and virtually all the manufacture until 1939.

France discovered that it had slipped behind in the gunmaking art when defeated by Germany in 1870, and in the years after that war the entire design and development facility was overhauled. As a result, the appearance of the M1897 75mm gun (see 'French 75') placed the French a good ten years ahead of everyone else. Unfortunately, it fitted too well with the current French tactical ideas of attacking and, as a result, far too much reliance was placed on it, the establishment of heavier guns and howitzers being neglected. The First World War showed the defects in this system and during that war some exceptionally good designs appeared, notably the 155mm GPF gun and the Schneider 155mm howitzer. Much early development work was done on self-propelled guns during the latter part of the war, though this was allowed to lapse when the war ended.

The French 75mm M1897 was the standard field-piece right up to the Second World War, even though by that time it was obsolescent. This picture shows a battery firing in pre-1914 days.

## FRENCH ARTILLERY EQUIPMENT

| Equipment | Date | Calibre (mm) | Barrel length (cals) | Breech mech. | Elevation max (deg) | Traverse (deg) | Weight in action (kg) | Shell weight (kg) | Muzzle velocity (m/sec) | Range, max (metres) |
|---|---|---|---|---|---|---|---|---|---|---|
| **Anti-tank artillery** | | | | | | | | | | |
| 25mm Hotchkiss Mle 34 | 1934 | 25 | 90 | VSB/SA | 15 | 60 | 480 | 0.32 | 918 | 1750 |
| | | | | | | | | | | |
| **Trench cannon** | | | | | | | | | | |
| 37mm M1916TR | 1916 | 37 | 22 | NS | 21 | 40 | 117 | 0.45 | 402 | 2400 |
| 45mm M1923 | 1923 | 45 | | NS | 45 | 7 | 198 | 1.8 | 450 | 5500 |
| | | | | | | | | | | |
| **Mountain artillery** | | | | | | | | | | |
| 65mm gun Mle 06 | 1908 | 65 | 20 | NS | 35 | 6 | 400 | 3.8 | 330 | 5500 |
| 70mm Schneider Mle 08 | 1908 | 70 | 17 | SQF | 20 | 4.5 | 508 | 5.3 | 300 | 5000 |
| 75mm Scheider | 1932 | 75 | | SQF | | | 400 | 6.5 | 220 | 4100 |
| 75mm Deport | 1910 | 75 | 17 | HSB/SA | 45 | 6 | 330 | 6.5 | 285 | |
| 75mm Mle 19 | 1919 | 75 | 18 | SQF | 40 | 10 | 659 | 6.5 | 450 | 9600 |
| 105mm How Mle 19 | 1919 | 105 | 12 | SQF | 43 | 9 | 750 | 12 | 350 | 7600 |
| | | | | | | | | | | |
| **Field artillery** | | | | | | | | | | |
| 75mm Gun Mle 1897 | 1897 | 75 | 36 | NS | 18 | 6 | 1138 | 7.2 | 530 | 6850 |
| 75mm Horse Arty Mle 12 | 1912 | 75 | 31 | SQF | 16 | 6 | 960 | 7.2 | 500 | 6500 |
| 75mm St Chamond | 1923 | 75 | | VSB | 70 | 40 | 75 | 3 | | 1800 |
| 105mm Gun Mle 13TR | 1913 | 105 | 22 | SQF | 37 | 6 | 1320 | 16.9 | 550 | 12700 |
| 105mm Gun Mle 36 | 1937 | 105 | 38 | HSB | 48 | 50 | 3540 | 15.7 | 725 | 16000 |
| 105mm How Mle 35 | 1937 | 105 | 17 | HSB | 50 | 58 | 1627 | 15.7 | 442 | 12000 |

In the 1920s, except for developing some highly automated guns for the Maginot Line, very little that was new appeared. The 75mm was overhauled and modernized by the addition of rubber tyres, and a useful 105mm howitzer was developed, but shortage of funds and general lethargy prevented these being produced in worthwhile numbers.

The Army was re-equipped with American weapons for several years after 1945 until it could take stock of its role and decide what was wanted. After that, however, design work went ahead and some excellent artillery has been produced in the last 25 years. The standard field-piece is still the US 105mm howitzer M1, though a new 105mm gun developed in 1984 may well replace that in the near future. Two 155mm howitzers, the 'TR' towed model and the 'GCT' self-propelled are now in full service as the standard general support weapons.

**The Schneider 155mm howitzer Mle 1917 became the principal medium howitzer, and was also adopted by the American Army.**

**Above: The 105mm Canon Court Mle 35B would have replaced the 75mm M1897 in due course, but by 1940 only 410 had been delivered. Notice that the wheels turned as the trails were opened to give additional protection.**

**Above: The Deport mountain howitzer of 1910 relied upon 'differential recoil' and is shown here with the gun in the forward position.**

**Left: A group of French gunners demonstrate 'abatage', a peculiarly French system of using the brake shoes as a form of firing segment. The gun is a 155mm Canon de Tir Rapide (CTR) Mle 1904.**

Left: The 105mm howitzer 'Mle 50 sur affût Automoteur' is the divisional close support SP howitzer, but will soon be replaced by the 155mm weapons.

Below: Details of the 155mm TR, showing the power-operated loading tray and rammer, and also showing how the wheels fold with the trail legs.

**Left:** The 155mm GCT (Grande Cadence de Tir) carries 40 rounds in the turret and uses automation to fire five rounds in 40 seconds.

**Below:** Canon de 194mm GPF mounted on a Saint-Chamond self-propelled chassis; developed in 1918, this was one of the earliest successful SP guns.

*French Artillery Equipment (Cont)*

| Equipment | Date | Calibre (mm) | Barrel Length (cals) | Breech mech. | Elevation max (deg) | Traverse (deg) | Weight in action (kg) | Shell weight (kg) | Muzzle velocity (m/sec) | Range max (metres) |
|---|---|---|---|---|---|---|---|---|---|---|
| **Heavy artillery** | | | | | | | | | | |
| 120mm Short Gun Mle 90 | 1890 | 120 | 14 | IS | 44 | 10 | 1475 | 18 | 290 | 5800 |
| 120mm How Mle 15TR | 1915 | 120 | 13 | IS | 43 | 5 | 1476 | 18.7 | 350 | 8300 |
| 139mm Gun Mle 10 | 1912 | 139 | 57 | IS | 30 | 6 | 11000 | 30.5 | 825 | 17400 |
| 145mm Gun Mle 16 | 1916 | 145 | 50 | IS | 38 | 6 | 12500 | 33.7 | 800 | 18500 |
| 155mm Gun Mle 77/14 | 1914 | 155 | 27 | IS | 42 | 5 | 6010 | 43 | 560 | 11400 |
| 155mm How Rimailho Mle 04 | 1905 | 155 | 15 | IS | 41 | 5 | 3200 | 40.3 | 320 | 6000 |
| 155mm How Mle 17 | 1917 | 155 | 15 | IS | 43 | 6 | 3300 | 43.5 | 450 | 11500 |
| 155mm How Mle 29 | 1930 | 155 | 22 | IS | 45 | 40 | 5165 | 38.6 | 635 | 15000 |
| 155mm How Mle 50 | 1952 | 155 | 23 | IS | 69 | 82 | 8155 | 43.7 | 645 | 17750 |
| 155mm Gun Mle 17LS | 1917 | 155 | 32 | IS | 40 | 5 | 8950 | 43.2 | 650 | 16000 |
| 155mm Gun Mle 17 GPF | 1917 | 155 | 38 | IS | 35 | 60 | 10750 | 43.1 | 735 | 16200 |
| 155mm Gun Mle 18 | 1918 | 155 | 27 | IS | 44 | 6 | 5030 | 43 | 560 | 12000 |
| 155mm Gun Mle 32 | 1933 | 155 | 50 | IS | 45 | 360 | 16400 | 50 | 900 | 26000 |
| 155mm Gun Mle TR | 1980 | 155 | 40 | HSB | 66 | 65 | 9500 | 43.2 | 810 | 23330 |
| 155mm SP Gun GCT | 1981 | 155 | | VSB | 66 | 360 | 38000 | 43.2 | 810 | 23330 |
| 22cm How Mle 01 | 1903 | 220 | 9 | IS | 40 | 40 | 8500 | 100 | | 7000 |
| 22cm How Mle 16 | 1915 | 220 | 10.3 | IS | 65 | 6 | 7800 | 100 | 415 | 11000 |
| 22cm Gun Mle 17 | 1917 | 220 | 28 | IS | 37 | 21 | 25000 | 103.5 | 765 | 22800 |
| 27cm How Mle 85 | 1885 | 270 | 9.6 | IS | 70 | 30 | 16500 | 152 | 326 | 8000 |
| 28cm Mortar Schneider | 1917 | 279 | 9.5 | IS | 60 | 20 | 16510 | 205 | 417 | 10900 |

The Schneider 90mm
Mle 1926
ready for
transport.

*French Artillery Equipment (Cont)*

| Equipment | Date | Calibre (mm) | Barrel Length (cals) | Breech mech. | Elevation max (deg) | Traverse (deg) | Weight in action (kg) | Shell weight (kg) | Muzzle velocity (m/sec) | Range max (metres) |
|---|---|---|---|---|---|---|---|---|---|---|
| **Railway artillery** | | | | | | | | | | |
| Matériel de 164 Mle93/96 | 1912 | 164.7 | 45 | | 40 | 360 | 57000 | 49.8 | 830 | 19200 |
| Matériel de 194 Mle 70/94 | 1914 | 194.4 | 24 | IS | 40 | 360 | 65000 | 83 | 830 | 13800 |
| Canon de 240 Mle 84 | 1915 | 240 | 21 | IS | 38 | 360 | 90000 | 159 | 575 | 17400 |
| Canon de 240 Mle 93/96 | 1915 | 240 | 31 | IS | 35 | 360 | 141000 | 162 | 840 | 22700 |
| Matériel de 274 Mle 87/93 | 1916 | 274 | 45 | IS | 40 | Nil | 152000 | 237 | 758 | 26400 |
| Canon de 285 Mle 17 | 1917 | 285 | 38 | | 40 | Nil | 152000 | 270 | 740 | 27400 |
| Matériel de 305 Mle 93/96 | 1915 | 305 | 30 | IS | 40 | Nil | 178000 | 348 | 795 | 27500 |
| Canon de 305 Mle 06 | 1916 | 305 | 38 | IS | 40 | Nil | 178000 | 388 | 674 | 28100 |
| Matériel de 305 Mle 06/10 | 1916 | 305 | 45 | IS | 38 | Nil | 208000 | 345 | 859 | 30800 |
| Matériel de 320 Mle 74 | 1915 | 320 | 30 | IS | 40 | Nil | 162000 | 388 | 675 | 24800 |
| Canon de 320 Mle 17 | 1917 | 320 | 35 | IS | 40 | Nil | 178000 | 392 | 690 | 28200 |
| Canon de 340 Mle 84 | 1917 | 340 | 28 | IS | 40 | 10 | 187000 | 452 | 580 | 18300 |
| Canon de 340 Mle 93 | 1917 | 340 | 35 | IS | 40 | 10 | 183000 | 462 | 740 | 26880 |
| Matériel de 340 Mle 12* | 1915 | 340 | 45 | IS | 37 | Nil | 270000 | 430 | 927 | 37600 |
| Matériel de 340 Mle 12** | 1915 | 340 | 45 | IS | 42 | 10 | 164000 | 432 | 920 | 44400 |
| Canon de 370 Mle 75/79 | 1915 | 340 | 24 | IS | 40 | Nil | 250000 | 709 | 600 | 24000 |
| Matériel de 370 Mle 15 | 1917 | 370 | 25 | IS | 65 | 12 | 130000 | 516 | 535 | 14600 |
| Matériel de 400 Mle 15/16 | 1917 | 400 | 25 | IS | 65 | 12 | 140000 | 641 | 530 | 16000 |
| Obusier de 520 | 1918 | 520 | 16 | IS | 60 | Nil | 260000 | 1654 | 450 | 14600 |

Notes: * Sliding mount    ** Rolling mount

## FRENCH SELF-PROPELLED ARTILLERY

| Equipment | Date | Calibre (mm) | Chassis | Elevation max (deg) | Traverse on mount (deg) | Weight in action (kg) | Range, max (metres) | Ammunition carried (rds) |
|---|---|---|---|---|---|---|---|---|
| AMX-13 DCA | 1964 | 30 twin | AMX-13 | 85 | 360 | 17200 | 3500 (AA) | 600 |
| F3 | 1952 | 155 | AMX-13 | 67 | 50 | 17400 | 20050 | nil |
| GCT | 1973 | 155 | AMX-30 | 66 | 360 | 42000 | 28500 | 42 |
| Modèle 50 | 1952 | 105 | AMX-13 | 66 | 40 | 16500 | 15000 | 56 |
| St Chamond | 1918 | 240 | Special | 35 | nil | 28000 | 17300 | nil |
| Schneider | 1918 | 220 | Special | 20 | nil | 26000 | 22800 | nil |

**FUME EXTRACTOR** A device fitted to the barrels of guns that have their breech in an enclosed space, such as a tank gun or turret-mounted gun, the purpose being to extact the poisonous fumes left in the bore after firing before the breech is opened and so prevent the fumes from entering the tank or turret. It consists of a short annular cylinder round the gun barrel, approximately two-thirds of the bore length from the chamber, and connected to the barrel by a port. This port is inclined rearwards from barrel to extractor cylinder, and as the projectile passes so the cylinder is filled with propellant gas at high pressure. Once the shell has left the muzzle and the pressure inside the barrel drops, the high-pressure gas inside the cylinder finds its escape through the port, which is sloping towards the muzzle. The rush of gas generates a flow through the bore which pulls out any remaining fumes.

The fume extractor (also called a 'bore eva-cuator') performs the same task as an air blast (qv), but does it in a much simpler manner.

**FUZE SETTER** A device used to set a time fuze attached to a shell before firing. The simplest form is a hand-held wrench, shaped to fit the adjustable section of the fuze. The gunner merely turns the fuze ring until a setting mark coincides with the required time marked on a scale on the body of the fuze. A more efficient system is to use a somewhat more complicated sort of wrench which carries a setting scale of greater precision than that engraved on the fuze. The time is set on the scale and the device is then dropped on to the nose of the fuze and rotated until it automatically stops when the fuze is correctly set.

A 'fuze-setting machine' was generally attached to the gun mounting, notably on anti-aircraft guns, so that the shell was pushed into the setter immediately before loading. A gunner operated this setter using a follow-the-pointer dial, keeping it always adjusted to the latest value as determined by the fire control system. The insertion of the fuze would automatically start a motor that caused the fuze to be correctly set.

An improvement on this system was to combine the fuze-setting machine with the mechanical loading system on more advanced designs, such as the British 3.7in and American 90mm anti-aircraft guns. In these the fuze setting data was automatically transmitted from the fire control predictor and did not require any attention from the gun detachment. The complete round (shell, fuze and cartridge) was simply dropped on to a loading tray, after which the fuze was automatically set, the round loaded and the gun fired.

An alternative method, experimented with by various nations with indifferent success, was to load the gun with the fuze set at safe and, by various forms of electric circuitry, set it while inside the gun or as it left the muzzle. This could be achieved by brushing the fuze with a wire as it left the barrel, so charging an electrical resistor-condensor circuit, or by setting magnets round the muzzle to affect soft iron cores inside the fuze.

Present-day technology is turning to electronic fuzes that can be set by simply applying a proportionate current or digital impulse to the fuze before loading. The more involved methods of automatic fuze setting and loading were rendered obsolete when the use of proximity fuzes for air defence became common.

**The operation of the fuze setter and automatic loading mechanism on the American 90mm M2 AA gun.**

FUNCTIONING OF FUZE SETTER-RAMMER

1.  Breech open, ramming rolls closed and rotating at low speed—fuze jaws closed.

2.  Round stationary, ramming rolls stalled and fuze jaws rotating fuze ring.

3.  Round jammed by ramming rolls rotating at high speed—fuze jaws open.

4.  Breech closed, gun is fired—in recoil ramming rolls open.

5.  Breech opened in counterrecoil, cartridge case ejected. Gun moves into battery, ramming rolls close and rotate at low speed—fuze jaws close.

# G

**GAMMA** Krupp began the development of large-calibre coast defence howitzers in the 1890s; the first was a 24cm weapon, called 'Alpha'; the next, in 1900, was a 30.5cm weapon called 'Beta'. Finally, in 1906, the company began development of a 42cm howitzer called 'Gamma', which was sufficiently powerful to destroy the strongest concrete fortifications likely to be encountered on Germany's borders. The weapon was completed by 1911 and firings took place which showed it to be phenomenally accurate to a range of 14.2 kilometres (8.8 miles), firing a shell weighing 1,150kg (2,535lb).

The howitzer weighed 140 tonnes when in the firing position, and it had to be dismantled by a crane and moved piecemeal on ten railway cars, taking about two days to be reassembled in its selected site. The German Army, while impressed with the performance, was less taken with the problem of moving it and asked for a fresh design which made the howitzer capable of being moved by road in fewer loads and assembled more quickly. This became 'Gamma M', more generally known as the 42cm Big Bertha (qv), and was adopted by the German Army and used against the Belgian and French forts in 1914.

'Gamma' remained at the Krupp proving ground at Meppen, close to the Dutch border, and was used as a test and experimental weapon, particularly for the development of concrete-piercing projectiles. It was dismantled in 1918 and appears to have escaped the attention of the Allied Disarmament Commission, so that in the early 1930s it was re-assembled and again put to work on experimental firings. A new shell was designed, weighing 1,003kg (2,211lb), and in 1942 Gamma was actually dismantled and shipped by rail to the vicinity of Sevastopol to add its weight to the siege. After that it was moved to Poland and fired into Warsaw during the 1944 rising.

**GARRINGTON GUN** An 87mm field gun designed by the Garrington Company in the 1950s as a potential replacement for the existing 25pr gun in the British Army. The gun iself was quite conventional, but the carriage was unusual; the box trail legs arched up from the axle to pass over the heads of the detachment before curving back to the ground. The upper part of this arch was covered by a thick glass-fibre 'umbrella' with side curtains, intended to protect the gun detachment from nuclear radiation and fallout. The gun appears to have passed its trials successfully, but the adoption of 105mm as the standard NATO close support gun calibre caused the project to be abruptly terminated.

The **87mm Garrington Gun**, showing the glass-fibre overhead shield intended to ward off shell splinters and nuclear radiation.

**GERLICH** A German engineer who developed a series of weapons in which the calibre gradually reduced from breech to muzzle, generally called the 'taper bore' (qv) system. The idea was originally patented by one Karl Puff in 1903, but lay fallow for many years. Gerlich resurrected the idea in the 1920s and developed a number of sporting rifles with moderate success. He then moved round the world, working for various agencies in Britain and the USA and attempting to promote the idea of taper-bore sniping rifle, though without success. He returned to Germany in the early 1930s; his subsequent movements are obscure, but both Rheinmetall-Borsig (qv) and Krupp (qv) later developed taper-bore guns based on his designs.

**The German 28/21mm Gerlich anti-tank gun undergoing trials in Britain in 1942.**

**The German 42cm Krupp howitzer in position on the Krupp experimental range at Meppen in the 1930s. The travelling gantry behind was used to assemble the weapon.**

The 105mm FlaK 39 anti-aircraft gun. Over 2000 defended Germany from 1938 to 1945.

Right: The 5cm PaK 38 anti-tank gun in action in Stalingrad.

**GERMAN ARTILLERY** Gun manufacture in Germany was principally in the hands of Krupp from the mid 19th century, though designs from Ehrhardt (working for what later became the Rheinmetall company [qv]) began to appear at the end of the century. The efficiency of German artillery in the Franco-Prussian War of 1870 made a considerable impression abroad and helped Krupp to build up an enormous export trade in artillery.

The German Army was outfitted with modern on-carriage recoil guns for its field artillery by World War One, and had numbers of heavier pieces. The war produced several new designs, notably in heavy artillery – 'Big Bertha', 'Max', the Paris Gun – but after the war the army was restricted to a handful of wartime designs. Work on new designs began in the 1920s, and surreptitious manufacture in about 1933, though all the new designs were given misleading model numbers (usually M18) to suggest that they were much older. In 1935 new guns began to appear in service, among them the 105mm 1eFH18, the standard divisional field piece; the 8.8cm FlaK18 anti-aircraft gun; the 15cm FH18, which became the backbone of German medium artillery; and numbers of heavy and railway guns were put into development. The Second World War saw the adoption of many new designs, from anti-tank guns with tapered bores to super-heavy weapons capable of firing shells of up to 7 tons in weight. It has to be said that several of the German guns were propaganda weapons, excellent for overawing enemies and impressing neutrals but of little practical value in the field. Nevertheless, there were some excellent designs in all classes, and development continued throughout the war with new weapons being introduced up until the early weeks of 1945.

The German Army was disbanded after the war and was not reconstituted as the Bundeswehr until the mid 1950s when it adopted American guns and howitzers almost exclusively. Since that time it has taken into service the tri-national FH70 howitzer, and has made some modifications to American and Italian weapons to better fit the German tactical doctrines.

This 88mm PaK 43/41 now sits beside an English country road, one of the few to survive the Russian front.

Inset: The 88mm PaK 43 was a purpose-built anti-tank gun, rather than a borrowed anti-aircraft gun, as had been the original 88s.

Right: Firing an 88mm FlaK 36 gun in the anti-tank role in the battle for Berlin in 1945.

| **GERMAN ARTILLERY EQUIPMENT** | | | | | | | | | |
|---|---|---|---|---|---|---|---|---|---|
| Equipment | Date | Calibre (mm) | Barrel length (cals) | Breech mech. | Elevation max (deg) | Traverse (deg) | Weight in action (kg) | Shell weight (kg) | Muzzle velocity (m/sec) | Range, max (metres) |
| **Anti-tank artillery** | | | | | | | | | | |
| 3.7cm PaK 1918 | 1918 | 37 | 22 | HSB | 9 | 21 | 175 | 0.6 | 345 | 2600 |
| 3.7cm PaK36 | 1936 | 37 | 42 | HSB | 25 | 60 | 432 | 0.7 | 762 | 4025 |
| 4.2cm LePaK41 Taper-bore | 1941 | 42/30 | 50 | HSB/SA | 32 | 60 | 450 | 0.3 | 1265 | 1000 |
| 5cm PaK38 | 1940 | 50 | 56 | HSB/SA | 27 | 65 | 986 | 2.2 | 823 | 2650 |
| 7.5cm PaK40 | 1941 | 75 | 42 | HSB | 22 | 65 | 1425 | 14.9 | 792 | 7680 |
| 7.5cm PaK41 Squeezebore | 1942 | 75/55 | 40 | HSB/SA | 17 | 60 | 1356 | 2.6 | 1125 | 4200 |
| 7.5cm PaK97/38 | 1941 | 75 | 33 | NS | 25 | 60 | 1190 | 4.5 | 570 | 1900 |
| 7.62cm PaK36(r) | 1942 | 76.2 | 38 | VSB/SA | 25 | 60 | 1730 | 7.5 | 740 | 9000 |
| 8cm PaW600 | 1944 | 81 | 35 | VSB | 32 | 55 | 600 | 2.7 | 520 | 750 |
| 8.8cm PaK43 | 1943 | 88 | 68 | VSB/SA | 40 | 360 | 3700 | 10.4 | 1000 | 17500 |
| 8.8cm PaK43/41 | 1943 | 88 | 68 | HSB/SA | 38 | 56 | 4380 | 10.4 | 1000 | 17500 |
| 12.8cm PaK44 | 1945 | 128 | 50 | HSB/SA | 45 | 360 | 10160 | 28.3 | 1000 | 24400 |

Left: The German Army used large numbers of guns under infantry control; this was one of the most common, the 75mm leIG (light infantry gun) 18 in use in the Donetz Basin in 1942.

Above right: German mountain troops with their 75mm GebG (mountain gun) 36.

Below: The Böhler-designed 105mm mountain howitzer, one of the best mountain weapons ever made. Many were still in use around Europe until the 1960s.

*German Artillery Equipment (Cont)*

| Equipment | Date | Calibre (mm) | Barrel Length (cals) | Breech mech. | Elevation max (deg) | Traverse (deg) | Weight in action (kg) | Shell weight (kg) | Muzzle velocity (m/sec) | Range max (metres) |
|---|---|---|---|---|---|---|---|---|---|---|
| **Anti-aircraft artillery** | | | | | | | | | | |
| 3.7cm FlaK 18,36 or 37 | 1937 | 37 | 85 | Auto | 85 | 360 | 1750 | 0.7 | 820 | 4800 ceiling |
| 5cm FlaK41 | 1941 | 50 | 68 | VSB/A | 90 | 360 | 3100 | 2.2 | 840 | 9000 ceiling |
| 7.5cm FlaK38 | 1928 | 75 | 60 | HSB | 85 | 360 | 3175 | 6.5 | 847 | 11500 ceiling |
| 7.7cm Ballonen-AK | 1914 | 77 | | HSB | 70 | 360 | 2500 | 7.65 | 510 | |
| 7.7cm Luftkanone | 1915 | 77 | 30 | NS | 80 | 360 | 1250 | 7.9 | 487 | |
| 8.8cm FlaK | 1918 | 88 | | HSB | 70 | 360 | 7300 | 15.3 | 785 | 3850 ceiling |
| 8.8cm FlaK 18,36 or 37 | 1934 | 88 | 53 | HSB/SA | 85 | 720 | 4985 | 9.4 | 820 | 9900 ceiling |
| 8.8cm FlaK41 | 1942 | 88 | 72 | HSB/SA | 90 | 360 | 7800 | 9.4 | 1000 | 15000 ceiling |
| 10.5cm FlaK 38 or 39 | 1939 | 105 | 53 | HSB/SA | 85 | 360 | 10224 | 14.8 | 881 | 11400 ceiling |
| 12.8cm FlaK40 | 1942 | 128 | 58 | HSB/SA | 88 | 360 | 27000 | 26 | 880 | 14800 ceiling |
| | | | | | | | | | | |
| **Mountain artillery** | | | | | | | | | | |
| 7.5cm Geb K08 | 1908 | 75 | 17 | HSB | 38 | 5 | 529 | 5.3 | 300 | 5750 |
| 7.5cm Geb G14 | 1914 | 75 | 16 | HSB | 36 | 5 | 491 | 5.3 | 280 | 4700 |
| 7.5cm Geb K15 | 1925 | 75 | 15 | HSB | 50 | 14 | 630 | 5.5 | 386 | 6625 |
| 7.5cm Geb G36 | 1938 | 75 | 19 | HSB | 70 | 40 | 750 | 5.7 | 475 | 9150 |
| 10.5cm Geb H L/12 | 1910 | 105 | 12 | HSB | 40 | 5 | 845 | 14.4 | 253 | 4900 |
| 10.5cm Geb H40 | 1942 | 105 | 32 | HSB | 71 | 51 | 1660 | 14.5 | 565 | 16740 |
| | | | | | | | | | | |
| **Infantry-accompanying artillery** | | | | | | | | | | |
| 7.5cm IG L/13 | 1936 | 75 | 13 | HSB | 43 | 50 | 376 | 6.3 | 225 | 3840 |
| 7.5cm IG18 | 1927 | 75 | 11 | VSB | 75 | 12 | 400 | 6 | 690 | 3375 |
| 7.5cm IG37 | 1944 | 75 | 20 | VSB/SA | 40 | 58 | 510 | 5.5 | 280 | 5150 |
| 7.5cm IG42 | 1944 | 75 | 20 | VSB/SA | 32 | 60 | 590 | 5.5 | 280 | 4600 |
| 15cm sIG33 | 1933 | 150 | 11 | HSB | 73 | 23 | 1700 | 38 | 240 | 4700 |

**Right:** The 75mm FK38 field gun was originally designed for sale to Brazil, but when war broke out in 1939 the order was diverted to the German Army.

**Below:** The biggest infantry-manned gun used by any army was this 15cm sIG (heavy infantry gun) 33. Self-propelled versions were also used.

In 1944 the German Army demanded some light but powerful dual-purpose anti-tank/ field guns. The 75mm FK 7M85 was the barrel of the 75mm PaK 40 fitted into the carriage of the 105mm leFH18/40. The result was a heavy anti-tank gun with support capability.

Left: The 75mm Mountain Gun M15 was designed by Skoda and became one of the most widely-used mountain weapons in Europe. It served in various armies from 1915 until 1945.

Right: The Krupp 120mm howitzer M1912, adopted by the German Army, during the 1913 manoeuvres.

Below: Standard divisional field howitzer of World War Two was this Rheinmetall-designed 105mm leFH 18.

*German Artillery Equipment (Cont)*

| Equipment | Date | Calibre (mm) | Barrel Length (cals) | Breech mech. | Elevation max (deg) | Traverse (deg) | Weight in action (kg) | Shell weight (kg) | Muzzle velocity (m/sec) | Range max (metres) |
|---|---|---|---|---|---|---|---|---|---|---|
| **Field artillery** | | | | | | | | | | |
| 7.5cm FK15nA | 1933 | 75 | 27 | HSB | 44 | 4 | 1525 | 5.8 | 662 | 12300 |
| 7.5cm leFK18 | 1938 | 75 | 22 | HSB | 45 | 30 | 1120 | 5.8 | 485 | 9425 |
| 7.5cm FK38 | 1942 | 75 | 25 | HSB/SA | 45 | 50 | 1365 | 5.8 | 605 | 11500 |
| 7.5cm FK7M85 | 1945 | 75 | 33 | HSB/SA | 42 | 30 | 1778 | 5.4 | 550 | 10275 |
| 7.7cm FK96 | 1896 | 77 | 27 | HSB | 16 | 4 | 1020 | 6.8 | 465 | 8465 |
| 7.7cm FK96nA | 1905 | 77 | 27 | HSB | 16 | 4 | 945 | 6.8 | 465 | 7800 |
| 7.7cm FK96/15 | 1915 | 77 | 27 | HSB | 15 | 4 | 1020 | 6.2 | 477 | 8400 |
| 7.7cm FK16 | 1916 | 77 | 35 | HSB | 40 | 4 | 1325 | 6.1 | 600 | 10300 |
| 10.5cm FH98/09 | 1909 | 105 | 16 | HSB | 40 | 4 | 1225 | 15.8 | 302 | 6300 |
| 10.5cm FH17 | 1917 | 105 | 20 | HSB | 43 | 4 | 1500 | 15.7 | 430 | 8950 |
| 10.5cm leFH16 | 1916 | 105 | 17 | HSB | 40 | 4 | 1450 | 14.8 | 395 | 9225 |
| 10.5cm leFH18 | 1935 | 105 | 24 | HSB | 40 | 56 | 1985 | 14.8 | 470 | 10675 |
| 10.5cm leFH18M | 1940 | 105 | 24 | HSB | 40 | 56 | 1985 | 14.2 | 540 | 12325 |
| 10.5cm leFH18/40 | 1941 | 105 | 25 | HSB | 40 | 56 | 1955 | 14.2 | 540 | 12325 |
| 10.5cm leFH42 | 1942 | 105 | 25 | HSB | 45 | 70 | 1630 | 14.8 | 595 | 13000 |
| 10.5cm leFH43 (Skoda) | 1944 | 105 | 23 | HSB | 75 | 360 | 2200 | 14.8 | 610 | 15000 |
| 10cm K17 | 1917 | 105 | | HSB | 45 | 6 | 3300 | 18.5 | 650 | 16500 |
| 13cm FK13 | 1913 | 135 | | HSB | | | | | | |

*German Artillery Equipment (Cont)*

| Equipment | Date | Calibre (mm) | Barrel Length (cals) | Breech mech. | Elevation max (deg) | Traverse (deg) | Weight in action (kg) | Shell weight (kg) | Muzzle velocity (m/sec) | Range max (metres) |
|---|---|---|---|---|---|---|---|---|---|---|
| **Heavy artillery** | | | | | | | | | | |
| s 10cm K18 | 1934 | 105 | 50 | HSB | 48 | 64 | 5642 | 15.1 | 835 | 19075 |
| 10.5cm sK18/40 | 1941 | 105 | 46 | HSB | 45 | 56 | 5680 | 15.1 | 885 | 20850 |
| 15cm sFH02 | 1902 | 150 | 11 | HSB | 65 | | 2189 | 39.5 | 276 | |
| 15cm sFH18 | 1934 | 150 | 27 | HSB | 45 | 64 | 5512 | 43.5 | 495 | 13250 |
| 15cm sFH13 | 1917 | 150 | 14 | HSB | 45 | 9 | 2250 | 42 | 381 | 8600 |
| 15cm sFH36 | 1938 | 150 | 16 | HSB | 45 | 56 | 3280 | 43.5 | 485 | 12300 |
| 15cm K16 | 1917 | 150 | | HSB | 43 | 8 | 10870 | 51.4 | 757 | 22000 |
| 15cm K18 | 1938 | 150 | 43 | HSB | 45 | 10 | 12760 | 43 | 890 | 24500 |
| 15cm K39 | 1940 | 149 | 53 | HSB | 46 | 60 | 12200 | 43 | 865 | 24700 |
| 15cm SKC/28 in Mörserlaf. | 1941 | 149 | 53 | HSB | 50 | 16 | 16870 | 43 | 890 | 23700 |
| 17cm K18 in Mörserlaf. | 1941 | 173 | 47 | HSB | 50 | 16 | 17520 | 62.8 | 925 | 29600 |
| 21cm Paris Gun | 1918 | 210 | 170 | HSB | 55 | 360 | 750000 | 119.7 | 1550 | 122300 |
| Lange 21cm Mörser | 1916 | 211 | 11 | HSB | 70 | 4 | 6680 | 113 | 393 | 11100 |
| 21cm Morser 18 | 1939 | 211 | 28 | HSB | 70 | 16 | 16700 | 133 | 565 | 16700 |
| 21cm K38 | 1943 | 211 | 52 | HSB | 50 | 17 | 25300 | 120 | 905 | 33900 |
| 21cm K39 | 1943 | 210 | 35 | IS | 45 | 360 | 33800 | 135 | 800 | 30000 |
| 24cm H39 | 1940 | 240 | 25 | IS | 70 | 360 | 27000 | 166 | 600 | 18000 |
| 24cm KL/46 | 1937 | 238 | 45 | HSB | 45 | 360 | 45200 | 180 | 850 | 32000 |
| 24cm K3 | 1938 | 240 | 52 | HSB | 56 | 360 | 54866 | 151.4 | 970 | 37500 |
| 28cm H L/12 | 1912 | 283 | 11 | HSB | 40 | 360 | 50300 | 350 | 350 | 10400 |
| 35.5 H M1 | 1939 | 356 | 22 | HSB | 75 | 360 | 75000 | 575 | 570 | 20000 |
| 42cm 'Gamma' H | 1906 | 420 | 12 | SQF | 75 | 46 | 140000 | 1003 | 452 | 14200 |
| 42cm Mrs L/14 'Big Bertha' | 1914 | 420 | 14 | HSB | 70 | 360 | 42600 | 816 | | 10250 |
| 54cm SP Mrs Karl | 1939 | 538 | 11 | HSB | 70 | 5 | 123000 | 1247 | 300 | 12500 |
| 60cm SP Mrs Karl | 1939 | 598 | 7 | HSB | 75 | 4 | 123000 | 1576 | 285 | 6675 |

Above: A 15cm sFH (heavy field howitzer) 18, the standard divisional medium howitzer, abandoned by its owners and regarded with curiosity by passing Soviet infantry.

Above left: The 21cm Schiffskanone L/40, a naval gun adopted for use in coast defence in the 1890s.

Right: The 28cm SK L/40 coast defence gun being prepared for action. Dating from 1901, this was an ex-naval gun put to shore use.

Below: A disappearing coast defence gun, the 15cm SK L/40 in the 1907 mounting, protecting the approaches to Kiel.

*German Artillery Equipment (Cont)*

| Equipment | Date | Calibre (mm) | Barrel Length (cals) | Breech mech. | Elevation max (deg) | Traverse (deg) | Weight in action (kg) | Shell weight (kg) | Muzzle velocity (m/sec) | Range max (metres) |
|---|---|---|---|---|---|---|---|---|---|---|
| **Railway artillery** | | | | | | | | | | |
| 15cm K in Eisenbahnlafette | 1937 | 149 | 37 | VSB | 45 | 360 | 75000 | 43 | 805 | 22500 |
| 17cm K (E) Samuel | 1916 | 173 | 40 | HSB | 47 | 12 | 60000 | 62.8 | 815 | 24020 |
| 17cm K (E) | 1938 | 173 | 28 | HSB | 45 | 360 | 80000 | 62.8 | 875 | 27200 |
| 20cm K (E) | 1941 | 203 | 57 | HSB | 47 | Nil | 86100 | 122 | 925 | 37000 |
| 21cm SKL/40 Peter Adalbert | 1916 | 209 | 40 | HSB | 45 | Nil | | 115 | 780 | 25580 |
| 21cm K12 (E) | 1939 | 211 | 152 | HSB | 55 | Nil | 302000 | 107.5 | 1500 | 115000 |
| 24cm K (E) Theodor Bruno | 1938 | 238 | 32 | HSB | 45 | 1 | 94000 | 148.5 | 675 | 20200 |
| 24cm K (E) Theodor | 1938 | 238 | 37 | HSB | 45 | 1 | 95000 | 148.5 | 810 | 26750 |
| 28cm K (E) Kurze Bruno | 1938 | 283 | 38 | HSB | 45 | 1 | 129000 | 240 | 820 | 29500 |
| 28cm K (E) Lange Bruno | 1937 | 283 | 40 | HSB | 45 | 1 | 123000 | 284 | 875 | 36100 |
| 28cm K (E) Schweret Bruno | 1938 | 283 | 40 | HSB | 45 | 1 | 118000 | 284 | 860 | 35700 |
| 28cm K (E) Neue Bruno | 1940 | 283 | 54 | HSB | 50 | 1 | 150000 | 265 | 995 | 36600 |
| 28cm K5 (E)* | 1940 | 283 | 72 | HSB | 50 | 1 | 218000 | 255 | 1128 | 62180 |
| 35cm K (E) | 1917 | 350 | 45 | HSB | 50 | 360 | 205000 | 700 | | 30000 |
| 38cm K (E) Siegfried | 1940 | 380 | 48 | HSB | 45 | Nil | 294000 | 495 | 1050 | 55700 |
| 40.6cm K (E) Adolf | 1940 | 406 | 48 | HSB | 45 | Nil | 323000 | 1030 | 610 | 56000 |
| 80cm K (E) Gustav | 1942 | 800 | 36 | HSB | 65 | Nil | 1350000 | 4800 | 820 | 47000 |
| | | | | | | | | | | |
| **Coast artillery** | | | | | | | | | | |
| 3.7cm SK C/30 | 1934 | 37 | 80 | VSB/SA | 80 | 260 | | 0.7 | 1000 | 6585 |
| 8.8cm SK C/35 | 1935 | 88 | 42 | VSB/SA | 30 | 360 | | | 700 | 12350 |
| 10.5cm SK C/32 | 1934 | 105 | 40 | VSB | 79 | 360 | 15231 | 15.1 | 785 | 15250 |
| 10.5cm SK L/60 | 1938 | 105 | 60 | VSB/SA | 80 | 360 | 11750 | 15.1 | 900 | 17500 |
| 15cm SK C/28 | 1936 | 149 | 52 | VSB/SA | 35 | 360 | | 45.5 | 785 | 23500 |
| 15cm TbtsK C/36 | 1937 | 149 | 45 | HSB | 40 | 120 | | 45.5 | 835 | 19525 |
| 15cm SK L/40 | 1914 | 149 | 40 | HSB | 30 | 360 | | 45.5 | 805 | 20000 |
| 15cm Ubts & Tbts K L/45 | 1934 | 150 | 45 | HSB | 45 | 360 | | 45.3 | 680 | 16000 |
| 17cm SK L/40 | 1908 | 173 | 40 | HSB | 45 | 360 | | 62.8 | 875 | 27200 |
| 20.3cm SK C/34 | 1934 | 203 | 57 | HSB | 40 | 360 | | 122 | 925 | 37000 |
| 24cm SK L/40 | 1914 | 240 | 40 | HSB | 45 | 360 | | 148.5 | 810 | 26750 |
| 24cm SK L/35 | 1910 | 240 | 35 | HSB | 45 | 360 | | 148.5 | 675 | 20200 |
| 28cm SK L/40 | 1901 | 283 | 40 | HSB | 45 | 360 | | 240 | 820 | 29500 |
| 28cm Küsten Haubitze | 1903 | 283 | 11 | HSB | 70 | 360 | 37000 | 350 | 379 | 11400 |
| 28cm SK L/45 | 1907 | 283 | 45 | HSB | 45 | 360 | | 284 | 875 | 36100 |
| 28cm SK L/50 | 1910 | 283 | 50 | HSB | 45 | 360 | | 284 | 905 | 39100 |
| 30.5cm SK L/50 | 1932 | 305 | 50 | HSB | 45 | 360 | 177000 | 250 | 1120 | 51000 |
| 35.6cm SK L/52.5 | 1914 | 356 | 52 | HSB | 52 | 360 | | 535 | | 50900 |
| 38cm SK L/45 'Max' | 1916 | 381 | 45 | HSB | 55 | 2 | 270000 | 400 | | 47500 |
| 40.6cm SK C/34 | 1939 | 406 | 45 | HSB | 60 | 360 | | 1030 | 610 | 56000 |

'Leopold' or 'Anzio Annie', the famous 28cm K5(E) railway gun, found its final resting place at Aberdeen Proving Ground in the USA.

The 10cm K41 field gun was devised in 1941 as a longer-ranging replacement for the 105mm howitzer but after much debate was not accepted.

The 15cm sFH 36 was intended to have the performance of the sFH18 but with a weight low enough to be hauled by a horse team. By extensive use of light alloy Rheinmetall produced this elegant weapon, but wartime shortages of alloy ended production in 1942.

Above: 'Hornisse', one of several tank destroyer designs carrying the 88mm anti-tank gun.

Right: The 21cm howitzer of 1916; fitted with an improved carriage it served well into World War Two.

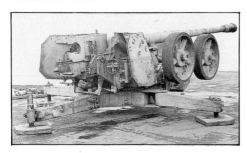

Above: Probably the most powerful anti-tank gun ever built, the Rheinmetall 12.8cm PaK 44 could penetrate 173mm of armour at 3km range and could throw a shell to 24km in the support role.

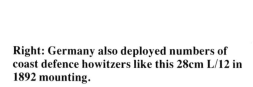

**Right: Germany also deployed numbers of coast defence howitzers like this 28cm L/12 in 1892 mounting.**

*German Artillery Equipment (Cont)*

| Equipment | Date | Calibre (mm) | Barrel Length (cals) | Breech mech. | Elevation max (deg) | Traverse (deg) | Weight in action (kg) | Shell weight (kg) | Muzzle velocity (m/sec) | Range max (metres) |
|---|---|---|---|---|---|---|---|---|---|---|
| **Recoilless artillery** | | | | | | | | | | |
| 7.5cm LG 40 | 1941 | 75 | 18 | HSB | 42 | 360 | 145 | 5.8 | 350 | 6800 |
| 7.5cm RFK 43 | 1944 | 75 | 9 | Special | 45 | 360 | 43 | 4 | 170 | 2000 |
| 10.5cm LG40 | 1941 | 105 | 13 | Special | 41 | 80 | 388 | 14.8 | 335 | 7950 |
| 10.5cm LG42 | 1941 | 105 | 13 | HSB | 42 | 360 | 540 | 14.8 | 835 | 7950 |

**Note:** * with standard pre-rifled shell; with rocket-assisted shell, 86.5km; with fin stabilized Arrow Shell, 151km.

## GERMAN SELF-PROPELLED ARTILLERY

| Equipment | Date | Calibre (mm) | Chassis | Elevation max (deg) | Traverse on mount (deg) | Weight in action (kg) | Range, max (metres) | Ammunition carried (rds) |
|---|---|---|---|---|---|---|---|---|
| 15cm sIG auf PzKwI | 1940 | 150 | Panzer I | 75 | 25 | 8500 | 4700 | 25 |
| 15cm sIG auf PzKwII | 1941 | 150 | Panzer II | 50 | 20 | 11200 | 4700 | 30 |
| 15cm sIG auf PzKw38(t) | 1943 | 150 | 38(t) tank | 72 | 10 | 11500 | 4700 | 15 |
| Brummbär | 1943 | 150 | Panzer IV | 30 | 20 | 28200 | 4700 | 38 |
| Elefant (Ferdinand) | 1943 | 88 | Tiger (P) | 14 | 28 | 65000 | 3000 (A/Tk) | 50 |
| Gepard | 1973 | 35 twin | Leopard | 85 | 360 | 47300 | 3500 (AA) | 620 |
| Grille | 1943 | 150 | 38(t) tank | 73 | 10 | 12000 | 4700 | 18 |
| Hetzer | 1944 | 75 | 38(t) tank | 12 | 16 | 17400 | 2000 (A/Tk) | 41 |
| Hummel | 1943 | 150 | Panzer IV | 42 | 30 | 23500 | 4700 | 18 |
| Jagdpanther | 1944 | 88 | Panther | 14 | 26 | 46000 | 3000 (A/Tk) | 57 |
| Jagtiger | 1944 | 128 | Tiger II | 15 | 20 | 70000 | 4000 (A/Tk) | 40 |
| Karl 040 | 1940 | 600 | Special | 70 | 8 | 123000 | 6675 | nil |
| Nashorn (Hornisse) | 1943 | 88 | Panzer IV | 20 | 30 | 23950 | 3000 (A/Tk) | 40 |
| Ostwind | 1944 | 37 | Panzer IV | 90 | 360 | 25000 | 4000 (AA) | 1000 |
| Sturmgeschütz IV | 1943 | 75 | Panzer IV | 15 | 20 | 24500 | 7675 | 79 |
| Sturmgeschütz 40 | 1940 | 75 | Panzer III | 20 | 24 | 22000 | 5000 | 44 |
| Wespe | 1942 | 105 | Panzer II | 45 | 34 | 11700 | 10675 | 32 |
| Wirbelwind | 1944 | 4×37 | Panzer IV | 90 | 360 | 22000 | 2000 (AA) | 3200 |

**'Granny', the British 9.2in howitzer dismantled into its basic units for transport behind a steam engine.**

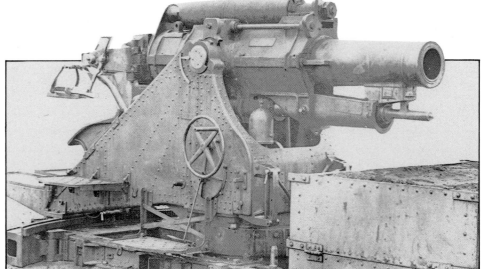

**Above: The British 9.2in howitzer of 1914. The steel box contained 11 tons of earth to anchor the mounting against recoil.**

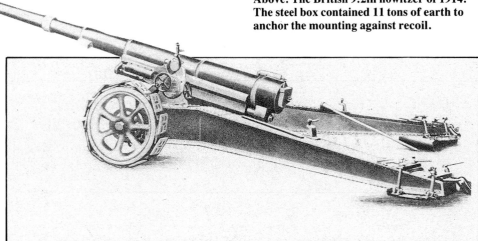

**GPF** An abbreviation meaning 'Grande Puissance, Filloux'; ie, gun of great power designed by Filloux. It was usually applied to the 155mm Gun Mle 1917, which was adopted by the French and US Armies during World War One, in order to identify it from other designs of 155mm gun of the period. The abbreviation was also applied to a 194mm gun designed by Filloux and adopted only by the French.

**GRANNY** The first 9.2in howitzer developed for the British Army in 1913-14. It completed its acceptance trials just as war broke out, and was sent to France forthwith; it acquired its nickname from the fact that it was the only one of its kind, and was bigger than any other British gun in the field army.

**GRAPHICAL FIRING TABLE** A specialized slide rule from which parameters such as elevation, fuze length, gun zone and time of flight can be determined by setting the specified range against an index. It was originally used as a quick method of determining gun data for simple targets; now principally held in fire direction centres as a back-up in case of computer failure.

**GRATICULE** The cross-wire or aiming mark in a telescope or sighting device; American usage is 'reticle' or 'reticule' which, strictly speaking, is not the same, being a pair of wires in a telescope used for determining the distance of an object.

**Left: The 155mm Grande Puissance Filloux (GPF) gun as adopted by the US Army from the French.**

**GRAVITY TANK** An oil tank attached to a gun carriage for the purpose of keeping the oil supply to the recoil buffer constantly topped up. Also called a 'replenisher', though strictly this name only applies when the oil is fed by pressure rather than purely by gravity.

**GREEN MACE** A British heavy anti-aircraft gun developed in the 1948-57 period. It was a 5 in calibre gun firing a fin-stabilized dart-like projectile and was designed by the Royal Armaments Research and Development Establishment (RARDE). Green Mace was fed from a rotating drum magazine, could fire at a rate of about 70 rounds per minute and had a water-cooled barrel. The design was successfully completed and prototypes built and fired, but when trials ended in 1957 it was almost immediately superseded by guided missiles. One drawback was the gun's weight, which was 28 tons, including its mobile mounting, and would have posed problems in deployment.

**GUN** Generally, a piece of artillery; specifically, an artillery piece that fires a projectile at high velocity with a relatively flat trajectory, always less than 45°, and uses a limited range of cartridge options. Compare with 'howitzer' (qv).

**GUNBUSTING** British gunner's term for what were officially called 'Repository Exercises', meaning the movement of ordnance and gun mountings, particularly coast defence and fortress guns, by the application of physical strength, aided by the use of levers, rollers, ropes, pulleys and simple hoists. Before the days of modern machinery, ordnance had to be emplaced by these methods, after it had been brought as close as possible to the position by boat, railway train, traction engine or horse-teams. These methods usually got the gun within a few hundred yards of its destined site, but the final yards could be difficult. In the case of a coast gun for a fort, it could mean hoisting it up flights of steps, through doors and passageways and finally into its casemate, after which the carriage had to be similarly brought up and the two fitted together in a confined space. Such 'shifts' could take days or weeks to accomplish, with as many as 100 men involved. The technique was maintained until the middle of the present century, because in many coast defence locations subsequent building made it impossible to bring mechanical handling equipment close to the gun, and the time-honoured methods had to be used if a barrel required changing or a new equipment brought in to replace an old.

**Installing a 10in gun in a British coast defence emplacement in 1896, using ropes, blocks, timber and plenty of manpower.**

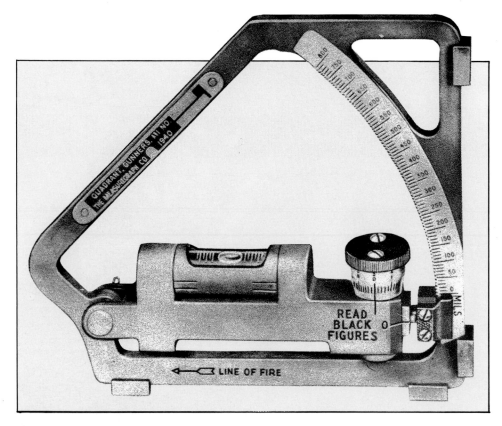

**A modern American gunner's quadrant used to apply the correct elevation to a gun barrel when the sight does not incorporate a level bubble.**

**GUN CORRECTIONS** Small alterations to the generally-ordered azimuth and range, intended to compensate for the gun's displacement from the theoretical battery centre, so that all the shells from the battery land in a straight and evenly-spaced line across the target. Also refers to corrections to individual guns that do not have calibrating sights (qv) to compensate for changes in muzzle velocity.

**GUN DRILL** Training exercises performed by the gun detachment in order to perfect their skill in serving the gun, but without firing. It is usually carried out on a parade square or gun park, but can be done at any convenient time or place.

**GUNFIRE** The fire for effect of a group of guns in which each gun loads and fires as rapidly as possible, without concern for the other guns of the group.

**GUN LIFT CARRIAGE** A type of gun mounting developed in the USA in the 1890s for coast defence. It consisted of a see-saw that carried a 12in gun at each end and was moved by a steam engine. With the see-saw horizontal, both guns were concealed behind the emplacement parapet. When required to fire, the steam engine tipped the see-saw so that one gun rose above the parapet and the other sank further behind the cover. After the first gun had fired, the see-saw was tipped in the other direction, raising the second gun to fire while the first was loaded behind cover. One installation was built at Fort Hancock, New Jersey, but proved to be a failure and was subsequently dismantled. The two gun mountings were modified to become barbette carriages and were installed in other forts.

**GUNNER'S QUADRANT** An instrument for measuring the elevation of the gun barrel. In ancient times it consisted of a straight-edge that was inserted into the gun muzzle and carried a scale graduated in 'points' at its front end, together with a plumb-bob. The straight-edge having been inserted into the barrel and held flush with the bottom of the bore, the plumb-bob hung down and cut across the scale; the gun was then elevated until the plumb-line cut the scale at the desired elevation.

The term is still used in the US and other armies, but now refers to the normal form of adjustable clinometer (qv) which can be placed on the gun breech, adjusted to the desired angle, and the gun elevated until the bubble on the instrument is level.

**An old engraving showing a gunner's quadrant in use.**

**A typical gunpit should be excavated, but is sometimes built up from sandbags (as here in Borneo) when the ground does not permit digging.**

**GUNPIT** Pit excavated in the ground to allow a gun to be concealed or protected from enemy fire. The term infers a field emplacement dug by the gun detachment, rather than a permanent emplacement lined with concrete.

**GUN RULE** A form of slide rule used to calculate fuze settings for use with shrapnel or smoke shells. The gun's muzzle velocity and the selected charge were set by cursors, the required range brought under the cursor index and the appropriate fuze length was then read opposite another index.

**GUSTAV** A German railway gun of 800mm calibre, formally known as the '80cm K(E)'. It was developed privately by Krupp in the late 1930s and then offered to the German Army in 1937 as a means of defeating the Maginot Line and similar fortifications. The idea was accepted, but manufacture proved to be more difficult than Krupp's had thought, and it was not finally fired at proof until 1941. After assembly and testing it was sent to Sevastopol and fired there during the siege. It was then sent to Leningrad, but before it could go into action the siege there was lifted. Gustav does not appear to have been used in any other combat, and most of it was found, in a wrecked condition, by American troops in Bavaria. It was scrapped after the war.

The gun was built in sections to enable it to pass the railway loading gauge. When assembled, the equipment was 42.79m (140ft) long, 7.01m (23ft) wide and 116m (38ft) high and weighed 1,350 tonnes (1,329 tons). The gun itself was dismantled into breech ring and block; barrel in two sections, jacket, cradle, trunnions and trunnion bearings, all carried on special flatcars. The mounting was split longitudinally for movement and dismantled from the top down, and the pieces were carried on additional special flatcars in a number of special trains. Assembly was done on a four-rail double track, with two outer tracks for the assembly crane. The right and left halves of the bogie units were placed in position and the gun carriage was built up on top. The barrel was assembled by inserting the rear half into the jacket, and connecting the front half by means of a massive junction nut and fitting the whole assembly to the cradle. The process of assembly took up to three weeks, and a force of 1,420 men commanded by a major-general was occupied in this task and in providing air defence and other services. Gustav fired a 4,800kg (4.7 ton) high explosive shell to 47 kilometres (29 miles) range or a 7,100kg (7 ton) concrete-piercing shell to 38 kilometres (24 miles). The 2,240kg (2.2 ton) charge was contained partly in bags and partly in a steel cartridge case 1.30m (4.3ft) long and 960mm (38in) in diameter.

Although not the largest calibre gun ever built, Gustav was certainly the largest artillery equipment in history. A second equipment, named 'Dora', is reputed to have been built, but there is little solid proof of this.

# H

**HANDSPIKE** A metal or metal-shod wooden pole used as a lever for lifting or moving guns or other heavy equipment.

**HANGFIRE** An ignition failure in a gun cartridge which results in a delay between operating the firing mechanism and the cartridge actually exploding. A hangfire is invariably due to faulty ignition of the cartridge. If a gun misfires it may be that a hangfire is in process, so the detachment will wait for a specified period of time before opening the breech to remove and examine the cartridge. The use of a washout (qv) obviates hangfire problems in heavy guns.

**HANGFIRE LATCH** An inertia-operated safety device fitted to the breech mechanism of some guns, which prevented the breech being opened unless the gun had fired or the catch was positively disengaged by a separate action. Found mostly on QF guns made prior to 1914 when the reliability of ammunition was not as good as it later became, it prevented an over-enthusiastic gunner opening the breech until the gun had fired, thus ensuring that he would not accidentally eject a burning cartridge that had hung fire.

**HARASSING FIRE** Artillery fire directed into rear areas at points of likely troop concentration, road junctions, suspected headquarters and similar targets and fired at irregular intervals to interrupt activities and lower morale.

**HEAVY ARTILLERY** A classification that covers guns and howitzers from about 155mm upwards; the qualifying calibre is not precise. In general, it means those pieces of artillery not organic to the division and controlled by the Corps or Army headquarters, and which are not expected to be deployed particularly rapidly. It includes guns and howitzers, and those weapons intended to counterattack enemy artillery. Railway artillery is also included in this classification by some countries, who qualify the term 'heavy' by 'road' or 'rail' nomenclature.

**Left: An American 155mm Gun M1 in use with British troops in Italy, 1944.**

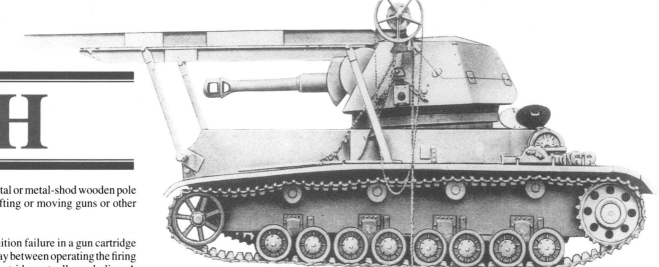

Above: The prototype Heuschrecke, showing how the turret was to be lifted off, carried forward and lowered on to a ground platform.

Below: A 19th century gun detachment at drill, showing the application of handspikes to move the carriage.

**HETZER** A German tank destroyer, formally known as 'Jagdpanzer 38(t)'. It was a tracked armoured vehicle based on the chassis of the Czech Model 38 tank, which had been taken over by the German Army on their occupation of that country. In place of the usual hull and turret there was a fixed superstructure mounting a 75mm gun in the front face of the vehicle, alongside the driver. Protected by 60mm (2.4in) of frontal armour, Hetzer carried a four-man crew and 41 rounds of ammunition. A total of 2,584 were manufactured between April 1944 and May 1945, and widely used on all fronts. The Swiss Army bought 150 after the war and continued to use them into the 1950s.

**HEUSCHRECKE** Heuschrecke ('Grasshopper') was a German self-propelled light field howitzer project, three prototypes of which were built in 1943. It consisted of the chassis of a Panzer IV tank

fitted with a removable turret mounting the standard leFH18 field howitzer. Along the sides of the tank were folding arms that carried a hoisting frame for erection above the hull. The turret could then be lifted from the hull by means of chains, moved to the rear of the tank, and lowered to the ground on to a prepared platform or on to a wheeled trailer. Once removed the turret could be emplaced as an armoured pillbox while the chassis was driven off to await the next move.

The idea was developed by Krupp, but the Army appears to have seen little virtue in it and the project was abandoned early in 1944.

**HIGH ANGLE FIRE** Artillery fire at angles of elevation over 45°; eg with howitzers. It is used in order to project the shell over intervening obstacles or to ensure a steep angle of descent at the target.

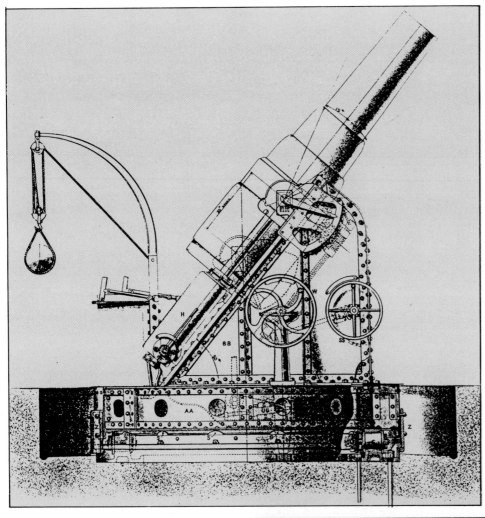

This objective was achieved by a gun with an extremely light barrel with a reinforced breech that fired a special round of ammunition. The cartridge case was closed by a heavy steel plate pierced with four small holes, and attached to the steel plate was the finned projectile. When loaded, the projectile lay in the smooth bore barrel, the cartridge case in the chamber, and the steel plate abutted against a step between chamber and barrel. When the cartridge was fired the propellant charge burned inside the case at very high pressure – about $1,250 \text{ kg/cm}^2$ ($17,780 \text{lb/in}^2$) – and the gas then passed through the holes in the steel plate to expand in the space behind the projectile. When the pressure in the barrel reached $400 \text{kg/cm}^2$ ($5,690 \text{lb/in}^2$) the projectile broke away from the cartridge case and was shot from the barrel at about $420 \text{m/sec}$ ($1,365 \text{ft/sec}$) to give it a maximum range of 6,200 metres (6,780 yards). Its accuracy at this range was not particularly good but at short ranges, as an anti-tank gun, it proved to be satisfactory. About 260 were made before the war ended.

The principle was examined after the war by other countries but has seen relatively little application; it is used by the British Royal Navy in an anti-submarine mortar, by the US Army in the M79 40mm grenade launcher, and by the Soviets in a 73mm gun used on infantry fighting vehicles.

**A British 9in rifled muzzle-loading gun on high angle mounting, for attacking the decks of armoured ships.**

**Below: The German 8cm Panzer Abwehr Werfer, the first high-low pressure gun; note the thin barrel and generally light construction.**

**HIGH ANGLE MOUNTING** Generally speaking, an old term for any gun mounting that allowed elevation greater than 45°. It was originally used for coast defence howitzers that fired at steep angles in order to drop shells on to the thinner decks of warships, instead of firing at flat trajectory against the thicker side armour. The term was then applied to the earliest anti-aircraft guns, but abandoned in this context at the time of the First World War when it was obvious that an anti-aircraft gun fired at high angles. It re-appeared in the 1920s with the development of coast defence gun mountings allowing elevations of 35°, as these differed from earlier mountings with maximum elevations of about 20°.

**HIGH-LOW PRESSURE GUN** Formally known as the 8cm Panzerabwehrwerfer 600, this was an anti-tank gun that introduced a totally new ballistic system. It was developed by Rheinmetall-Borsig (qv) in 1943 in response to a German Army demand for an anti-tank gun light enough for two or three infantrymen to manhandle, that used less propellant than a recoilless gun or a rocket and yet would be accurate enough to hit a 1m square target at 750 metres (820 yards) range.

**HIGH PRESSURE PUMP** A German 15cm long-range gun known as the 'Hochdruckpumpe' (HDP) and also as 'Busy Lizzie', 'The Centipede' and Vergeltungswaffe 3.

The HDP was a further manifestation of the multiple-chamber gun (qv), which began as far back as the 1880s with the Lyman and Haskell gun (qv). The principle is to have a long gun barrel with a conventional chamber and with additional chambers arranged herringbone fashion along the bore. The shell is fired by the charge in the conventional chambers, and the charges in the side chambers are fired as the shell passes, so adding more propelling gas behind the shell. As a result the shell increases in velocity and develops a very long range.

The idea for the HDP came from an engineer named Conders, who worked for the Röchling Steel Company in Germany. This firm had developed a fin-stabilized shell for the defeat of concrete fortifications, and it is probable that Conders envisaged this gun originally as a better method of propelling this shell to long ranges. In May 1943 he manufactured a 20mm scale model of the gun which appeared to work well. He sent a report to Reichsminister Speer, Hitler's munitions production chief, and through him the report reached Hitler. The proposal was for a battery of 50 15cm HDP barrels to be built into a hillside near Mimoyècques, France, and aimed to bombard London. This grandiose idea captured Hitler's imagination and orders were given for construction of the emplacement to begin, even though a working gun had not yet been made.

A short 15cm gun had been installed at the Hillersleben Proving Ground in Germany by September 1943. The original idea was to fire the side chambers electrically, but this was abandoned and the flash of the propellant fired each charge as the shell went past the chamber. This gave rise to the same trouble which had dogged the Lyman and Haskell gun, that of flashover past the shell igniting the side charges in advance of the shell and setting up back-pressures. The fault was cured by putting a highly efficient sealing piston behind the projectile.

In 1944 a full-length 150 metre (492 foot) barrel was assembled at Misdroy, on the Baltic coast, and firing trials began, together with training for the men destined for the Momoyècques battery. The shells were unstable and had to be redesigned, and trials were often interrupted because sections of barrel were blown out by abnormally high pressures within the gun. By May 1944 most of the technical problems appeared to have been solved, and the predicted maximum range of 150 kilometres (93.2 miles) was in sight. Meanwhile the V1 flying bomb campaign had begun, and in retaliation British and US aircraft ranged over France seeking out anything that might be a flying bomb launch site; they discovered Mimoyècques and wrecked it. Shortly afterwards the Allies invaded France, and before the firing trials were completed the Mimoyècques site had been overrun and finally destroyed by blasting.

Two short versions of the 15cm HDP were

The 15cm 'High Pressure Pump' long range gun laid out for testing at the Hillersleben Artillery Proving Ground in Germany.

eventually assembled and sent into action. One was mounted on a railway flatcar and fired into Antwerp, while the other was laid out on a hillside near Hermeskeil and fired into Luxembourg. Only a few shots were fired before the Allied advance made it necessary to destroy the guns and retreat. No data have ever been discovered about the performance of these two weapons. The planned data for the Mimoyècques type gun were that the barrels would be 150m (492ft) long and fixed at an elevation of 55°; they would fire an 83kg (183lb) shell at a muzzle velocity of 1,463m/sec (4,800ft/sec) to achieve a range of 150 kilometres (93.2 miles).

A British 12in Mark 4 howitzer. The steel box on the left contained 15 tons of earth to anchor the carriage when the weapon fired.

**HOLDFAST** An anchorage used with guns of position; ie, coast defence, anti-aircraft or siege guns. It usually consisted of a steel ring set into the concrete floor of the emplacement and carrying a number of bolts to which the gun mounting was attached. In portable siege guns it was usually a series of ground anchors and cables to secure the gun to a wooden platform.

**HOOPS** American terminology for the various tubes or layers in a built-up gun which, shrunk upon one another, place the innermost tube in the desired state of compression to resist the internal pressure on firing.

**HORNISSE ('HORNET')** German tank destroyer mounting an 88mm gun on the chassis of the Panzer IV tank. The normal hull and turret of the tank were removed and an open-topped superstructure built up, carrying the gun in a limited-traverse mounting on the front plate. Development began in 1942, the aim being to provide a self-propelled mounting for the forthcoming PaK43 anti-tank gun in time for the 1943 summer offensive. Production began in February 1943 and 494 were built before the war ended. Lightly armoured, it was nimble and carried an extremely potent gun with the result that it became a highly effective tank-killer. It was also known as 'Nashorn' (Rhinoceros).

**HORSE ARTILLERY** Field artillery of a light nature, allotted to the support of cavalry formations. In the days when all artillery was horse-drawn, the distinction attaching to Horse Artillery was principally that every man of the gun detachment rode a horse, whereas in field artillery men might march alongside the gun or ride on the gun, on the limber or in an accompanying wagon. The guns were usually of 75mm calibre and as light as possible consistent with achieving a reasonable range. The object was to have an artillery unit which could keep up with the cavalry in advance.

Horse artillery survived the arrival of mechanization in the British Army with the role of field artillery in support of an armoured division, being armed with the same weapon as the field artillery, though in general Royal Horse Artillery regiments were more likely to be equipped with self-propelled guns.

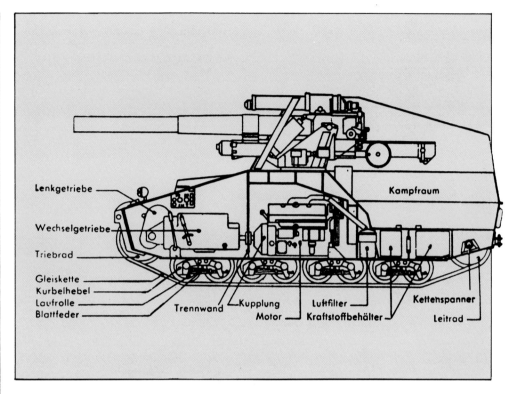

A design showing the principal parts of the German Hummel SP howitzer, which was one of the few German self-propelled guns to be simply a mobile field artillery piece and not an assault gun. It carries the 15cm heavy howitzer.

**HOTCHKISS** A French company founded by an American, Benjamin Berkeley Hotchkiss, in the 1870s and more associated with small arms than with artillery for most of its existence. However, in the 1880s it developed a breech-loading 3pr 47mm gun that used a fixed round of ammunition (a round in which the shell was attached to the cartridge case like an overgrown rifle cartridge) and was widely adopted by navies as an anti-torpedo-boat gun. It used a vertical sliding breechblock and is generally considered to be the earliest quick-firing gun of modern appearance, though it did not use any form of recoil system. This was followed by a 6pr of 57mm calibre along similar lines. Both were used in British coast defences as anti-torpedo-boat guns, though the 3pr was soon relegated to a practice and drill weapon. Hotchkiss also promoted a 37mm revolving-barrel cannon, which was similar in some respects to the Gatling gun and was adopted in several navies to repel boarders.

**HOWITZER** A piece of artillery capable of firing at angles above 45° and at a low velocity. It uses a propelling charge capable of being adjusted for size to give a number of trajectory options, all of which describe a high path to pass over intervening obstacles and reach behind cover to drop the shell on to the target. Compared with a gun of the same calibre the howitzer fires a heavier shell at lower velocity and has a shorter range.

**HUMMEL** A German self-propelled 15cm heavy howitzer mounted on a lengthened Panzer IV hull and using some components of the Panzer III tank. The hull was built up into a high open-topped box that had the 15cm sFH18 howitzer mounted centrally and protruding through a slot in the front plate. The vehicle was crewed by five men and a driver, and carried 18 complete rounds of ammunition, further ammunition being brought up by another converted tank acting as a munitions carrier. About 100 guns and 150 munitions carriers were built and issued to artillery in support of armoured formations, notably on the Russian front. The gun had 15° of traverse each side of centre, and had a maximum elevation of 42°, which gave it a range of 12.5 kilometres (7.8 miles) with a 43.5kg (96lb) shell.

**HYDRO-PNEUMATIC** A mechanism relying upon the action of a liquid and a compressed gas; in artillery terms, usually refers to either a recoil system or to an equilibrator (qv). In the case of a recoil system the recoil is controlled by the liquid (see 'buffer') and the gun is then run out by the action of the gas, which has been compressed during the recoil movement.

**HYDRO-SPRING** A mechanism relying upon the action of a liquid and a spring; in artillery terms it refers always to a recoil system in which recoil is controlled by the liquid and the run-out is performed by springs.

The American 105mm howitzer M3, showing the two sets of sights required by the independent system. The azimuth sights are visible on the left of the gun, and the elevation sights on the right.

# I

**INDEPENDENT LINE OF SIGHT** A system of gunlaying in which the elevation of the sight and the elevation of the gun barrel are controlled independently. Thus the gunlayer lays his sight on the target for azimuth, while a second gunlayer, generally on the other side of the gun, lays the barrel for elevation by means of a clinometer (qv). It allows changes of elevation to be made without moving the azimuth sight off the target. Developed originally by the French for their 75mm gun of 1897, it is only of value in direct-fire guns, and once field guns began to be used principally in the indirect-fire role the independent line of sight system was gradually abandoned. It has also been called 'split function laying'.

**INDIRECT FIRE** Artillery fire is called 'indirect' when the gun and the target are not intervisible. It is the majority case except for anti-tank, anti-aircraft and most types of coast defence artillery. Indirect fire requires a forward observer to spot targets, observe the fall of shot and correct it if necessary, and it requires a system of sights that does not rely on seeing the target. The most usual system is to select an 'aiming point' that is conveniently visible to all the guns, calculate the azimuth from the centre of the gun battery to this point; calculate the azimuth to the target and determine the angle between target and aiming point; set this angle on the sight; then move the gun until the sight points at the aiming point, when the gun barrel will be pointing at the target. The range is applied by looking up a table of ranges and elevations, selecting the required elevation, and applying it to the gun barrel by means of a clinometer (qv).

**INTENSIFIER** A form of hydraulic seal applied to the recoil system piston rods as they leave their cylinders, to prevent leakage of oil. It is used only with heavy equipments in static mountings (eg heavy anti-aircraft and coast defence guns) where the recoil is severely curtailed in the interests of a high rate of fire and therefore the pressure generated inside the recoil system is high. Ordinary gland packings and seals tend to develop leaks under these conditions, but the hydraulic intensifier developed a higher sealing pressure in proportion to the rise of pressure inside the cylinder.

**INTERDICTION** Loose term used to describe fire against targets in rear areas, intended to deny or inhibit the use of an area or to attack echelons following an assault. Also used as an alternative name for 'harassing fire' (qv).

**ISRAELI ARTILLERY** When the Israeli Defence Force was formed in 1948 its artillery was armed with a varied collection of equipment from numerous sources. In time, standardization on American equipment, notably the 105mm Howitzer M1 and the 155mm Howitzer M114, was achieved, but the force has since been augmented by large amounts of Soviet equipments captured from Arab League countries at various times. During the 1960s the Soltam company, which has connections with the Tampella company of Finland, took a Finnish design of 155mm howitzer and made modifications to suit the requirements of the IDF. This design, the M68, has since been improved into M71, Model 839P and Model 845P variants, some in use in Israel and others exported and used by Singapore and Thailand.

The Israelis have also developed a number of indigenous self-propelled equipments adapted from existing American or Soviet guns which have been married to various redundant tank chassis that have become available as more modern tanks have been put into service.

## ISRAELI ARTILLERY EQUIPMENT

| Equipment | Date | Calibre (mm) | Barrel length (cals) | Breech mech. | Elevation max (deg) | Traverse (deg) | Weight in action (kg) | Shell weight (kg) | Muzzle velocity (m/sec) | Range, max (metres) |
|---|---|---|---|---|---|---|---|---|---|---|
| Soltam M-68 Gun-howitzer | 1968 | 155 | 30 | HSB | 52 | 90 | 8500 | 44 | 725 | 21000 |
| Soltam M-71 Gun-howitzer | 1971 | 155 | 39 | HSB | 55 | 90 | 9000 | 44 | 765 | 23000 |
| Soltam Model 839P | 1985 | 155 | 39 | HSB | 70 | 78 | 10850 | 44 | | 31000 |
| Soltam Model 845P | 1985 | 155 | 45 | HSB | 70 | 78 | 11700 | 44 | | 39000 |

## ISRAELI SELF-PROPELLED ARTILLERY

| Equipment | Date | Calibre (mm) | Chassis | Elevation max (deg) | Traverse on mount (deg) | Weight in action (kg) | Range, max (metres) | Ammunition carried (rds) |
|---|---|---|---|---|---|---|---|---|
| M50 | 1958 | 155 | Sherman | 69 | 40 | 31000 | 17600 | |
| Soltam L33 | 1967 | 155 | Sherman | 52 | 60 | 41500 | 20000 | 60 |
| Soltam M72 (prototype) | 1985 | 155 | Centurion | 65 | 360 | | 23500 | |

**ITALIAN ARTILLERY** The Italian Army was almost entirely equipped with guns purchased from abroad prior to 1918, principally from Krupp, Skoda or Deport of France. The only exception to this was a small number of mountain artillery guns developed and manufactured in Turin Arsenal. Some foreign guns were manufactured under license by Vickers-Terni. More Skoda designs were acquired after 1918, as reparations from Austria, and in the 1930s the Ansaldo company began to design and manufacture guns of purely Italian origin. Their production was insufficient to replace the earlier guns, however, and these remained in use throughout the Second World War.

British guns were used for some years after 1945, but the artillery was later standardized on American models. In the 1950s the OTO-Melara company developed the 105mm M56 pack howitzer, a highly successful design which was widely adopted throughout the world, and Italy collaborated in the development of the 155mm FH70 howitzer.

The 100mm '100/17' field gun Model 1914, designed by Skoda.

## ITALIAN ARTILLERY EQUIPMENT

| Equipment | Date | Calibre (mm) | Barrel length (cals) | Breech mech. | Elevation max (deg) | Traverse (deg) | Weight in action (kg) | Shell weight (kg) | Muzzle velocity (m/sec) | Range, max (metres) |
|---|---|---|---|---|---|---|---|---|---|---|
| **Anti-tank artillery** | | | | | | | | | | |
| Canone da 47/32 Mo. 35 | 1935 | 47 | 36 | HSB | 58 | 60 | 265 | 1.5 | 630 | 8200 |
| Canone da 47/32 Mo. 39 | 1939 | 47 | 36 | HSB | 56 | 60 | 277 | 1.5 | 630 | 7000 |

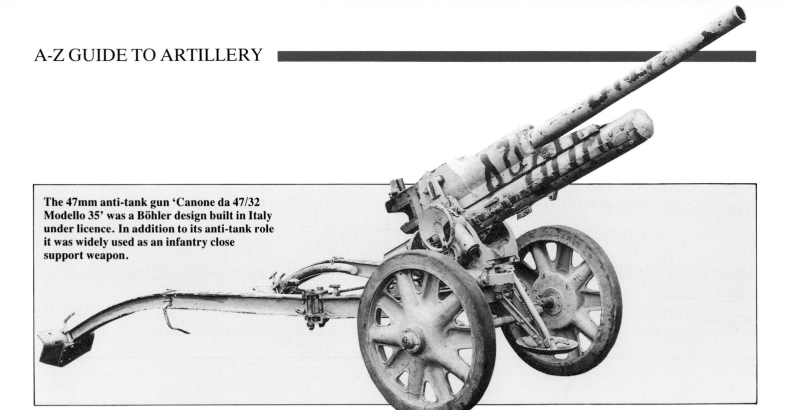

The 47mm anti-tank gun 'Canone da 47/32 Modello 35' was a Böhler design built in Italy under licence. In addition to its anti-tank role it was widely used as an infantry close support weapon.

*Israeli Artillery Equipment (Cont)*

| Equipment | Date | Calibre (mm) | Barrel Length (cals) | Breech mech. | Elevation max (deg) | Traverse (deg) | Weight in action (kg) | Shell weight (kg) | Muzzle velocity (m/sec) | Range max (metres) |
|---|---|---|---|---|---|---|---|---|---|---|
| **Anti-aircraft artillery** | | | | | | | | | | |
| Breda 37/54 | 1925 | 37 | 54 | VSB | 90 | 360 | | 0.79 | 800 | 4200 ceiling |
| 75/27 Mo. 06/15 | 1915 | 75 | 27 | HSB/SA | 70 | 360 | 4675 | 6.5 | 510 | 4600 ceiling |
| 75/46 Mo. 35 Ansaldo | 1935 | 75 | 46 | HSB/SA | 90 | 360 | 3350 | 6.5 | 715 | 9300 ceiling |
| 75/49 Skoda | 1935 ? | 75 | 49 | VSB | 85 | 360 | 2850 | 6.3 | 820 | 9150 ceiling |
| 75/50 Ansaldo | 1938 | 75 | 50 | HSB/SA | 90 | 360 | 5200 | 6.5 | 975 | 8400 ceiling |
| 76/40 Ansaldo | 1912 | 76.2 | 40 | SQF | 80 | 360 | | 6.5 | 690 | 5000 ceiling |
| 76/45 Mo. 11 | 1912 | 76.2 | 45 | SQF | 80 | 360 | 2235 | 6.3 | 750 | 6400 ceiling |
| 90/53 Ansaldo | 1938 | 90 | 53 | HSB/SA | 85 | 360 | 5180 | 10 | 840 | 12000 ceiling |
| 102/35 | 1938 | 102 | 35 | VSB/SA | 70 | 360 | | 13.2 | 755 | 9500 ceiling |
| | | | | | | | | | | |
| **Mountain artillery** | | | | | | | | | | |
| Canone da 65/17 | 1913 | 65 | 17 | SQF | 20 | 8 | 556 | 4.3 | 345 | 6800 |
| 70mm Mo. 02 | 1903 | 70 | 16 | SQF | 21 | Nil | 387 | 4.8 | 353 | 6630 |
| 70mm Mo. 08 | 1910 | 70 | 17 | HSB | 20 | 5 | 508 | 5.3 | 300 | 5000 |
| Canone da 70/15 Turin | | 70 | 16 | SQF | 21 | Nil | 387 | 4.7 | 353 | 6600 |
| Canone da 75/13 Skoda | 1922 | 75 | 13 | HSB | | | | | | |
| Obice da 75/18 Mo. 34 | 1935 | 75 | 18 | HSB | 65 | 48 | 800 | 6.3 | 435 | 9400 |
| | | | | | | | | | | |
| **Field artillery** | | | | | | | | | | |
| 75/27 Mo. '06 | 1907 | 75 | 27 | HSB | 17 | 7 | 1005 | 6.5 | 510 | 6800 |
| 75mm Gun Mo. 06/12 | 1913 | 75 | 30 | HSB | 19 | 7 | 958 | 6.5 | 510 | 7600 |
| 75mm Gun Mo. 12 Deport | 1912 | 75 | 27 | NS | 65 | 52 | 1076 | 6.5 | 510 | 7600 |
| 75/32 Mo. 37 | 1937 | 75 | 32 | HSB | 45 | 50 | 1185 | 6.3 | 600 | 12500 |
| Obice da 75/18 Mo. 35 | 1935 | 75 | 18 | HSB | 45 | 50 | 1100 | 6.35 | 435 | 9400 |
| Obice da 100/17 Mo. 14 | 1914 | 100 | 17 | SQF | 48 | 5 | 1450 | 12.5 | 410 | 9260 |
| Obice da 105/28 | | 105 | 28 | HSB | 37 | 14 | 2470 | 15.5 | 570 | 13200 |
| Canone da 105/40 Mo. 42 | 1942 | 105 | 40 | HSB | 45 | 50 | 3860 | 17.5 | 710 | 17600 |
| Obice da 105/14 Mo. 56 | 1956 | 105 | 14 | VSB/SA | 65 | 56 | 1273 | 14.9 | 416 | 11100 |

The 75mm OTO-Melara M56 pack howitzer, widely adopted by several armies, seen here with the trail legs folded up and fitted with a towing attachment.

The 75mm M56 at high elevation and with the trails set in the high position. They can be set in a lower position to give better stability when firing in the anti-tank role.

Above: The 75mm 'Obice da 75/18' used the same carriage as the Model 37 field gun but had a shorter barrel.

Left: The 105/32 field howitzer, the standard Italian field gun during the Second World War.

Above: The Ansaldo 149mm heavy field gun Model 35, another modern and sound design of which the army never had sufficient.

Above right: The 75mm anti-aircraft gun 'Canone de 75/46 Modello 34' in use in the ground role in Libya, 1941.

Left: 21cm howitzer 'Mortaio de 210/8', model of 1910, at Val Dogno during World War One, firing against the Austrians.

Front view of the 75mm Model 37, showing the unusual perforated jacket around the barrel and the conical muzzle brake.

*Italian Artillery Equipment (Cont)*

| Equipment | Date | Calibre (mm) | Barrel Length (cals) | Breech mech. | Elevation max (deg) | Traverse (deg) | Weight in action (kg) | Shell weight (kg) | Muzzle velocity (m/sec) | Range max (metres) |
|---|---|---|---|---|---|---|---|---|---|---|
| **Heavy artillery** | | | | | | | | | | |
| Obice da 149/12 | | 149 | 12 | HSB | 65 | 5 | 2390 | 41 | 300 | 6600 |
| Obice da 149/13 M14/15 | 1915 | 149 | 13 | HSB | 70 | 6 | 2795 | 42.2 | 345 | 8750 |
| Canone da 149/35 | 1910 | 149 | 35 | IS | 35 | Nil | 8200 | 45.8 | 700 | 9700 |
| Canone da 149/40 Mo. 35 | 1935 | 149 | 40 | IS | 45 | 60 | 11480 | 50.8 | 800 | 22000 |
| Canone da 152/37 Mo. 15 | 1915 | 152.4 | 37 | IS | 45 | 6 | 11890 | 54 | 700 | 24500 |
| Canone da 155/25 | 1908 | 155 | 25 | IS | 28 | Nil | 7180 | 43 | 500 | 11400 |
| Mortaio da 210/8 | | 210 | 8 | IS | 70 | 360 | 10935 | 100.6 | 370 | 8000 |
| Obice da 210/22 Mo. 35 | 1934 | 210 | 22 | IS | 70 | 30 | 15800 | 102 | 570 | 16000 |
| Mortaio da 260/9 Mo. 16 | 1916 | 260 | 9 | IS | 65 | 30 | 11830 | 219 | 300 | 9100 |
| Mortaio da 305/8 Mo. 11/16 | 1916 | 305 | 8 | IS | 75 | 120 | 20880 | 380 | 400 | 11000 |
| Obice da 305/17DS | 1915 | 305 | 17 | IS | 65 | | 34290 | 350 | 545 | 17550 |

## ITALIAN SELF-PROPELLED ARTILLERY

| Equipment | Date | Calibre (mm) | Chassis | Elevation max (deg) | Traverse on mount (deg) | Weight in action (kg) | Range, max (metres) | Ammunition carried (rds) |
|---|---|---|---|---|---|---|---|---|
| OTO-76 | 1984 | 76 | Palmaria | 60 | 360 | 50000 | 6000 (AA) | 90 |
| Palmaria | 1982 | 155 | OF-40 tank | 70 | 360 | 46000 | 27500 | 30 |
| Semovente 90 | 1942 | 90 | M41 tank | 24 | 80 | 17000 | 19000 | 6 |
| Semovente 149 | 1942 | 149 | M13/40 tank | 45 | 60 | 18000 | 23500 | 6 |

# J

**JACKET** The outer tube round all or part of a gun barrel and which supports and protects the barrel and supplies a means of attachment to the cradle. It assists in giving longitudinal and girder strength to the gun, but plays no part in resisting internal pressure.

**JANACEK** A Czech gun designer who escaped from Czechoslovakia at the time of the German occupation, Janaček came to Britain and worked with the BSA company on various weapons. He proposed that a taper-bore adaptor (qv) be fitted to the muzzle of the 40mm 2pr gun mounted in armoured cars and tanks. Used with a special tungsten-cored projectile that could squeeze down in calibre as it went through the adaptor, this proposal would give the gun an extremely high velocity and extend its useful life against the heavier types of tank which, using conventional ammunition, it could no longer damage. Technically speaking this 'Littlejohn Adaptor' ('Littlejohn' being the Anglicized version of 'Janaček') was a success, but it took so long to perfect that by the time it reached service the 2pr gun was on its way out. It was used on some British and American armoured cars and light tanks with 40mm and 37mm guns.

**JAPANESE ARTILLERY** There was no Japanese industry capable of manufacturing artillery prior to 1914, and all guns were therefore purchased from foreign firms, principally from Krupp. Work began during the war to build foreign guns with and without licence, and also modify earlier guns to meet Japanese Army requirements. Osaka Arsenal became the principal gun design and manufacturing agency and, after 1918, produced a wide range of weapons from pack artillery to heavy coast defence guns. In general, Japanese field artillery was always notable for its long range and light weight in comparison with European designs of similar calibre and employment.

The Japanese Army standardized on American equipment after the Second World War and still does, though it developed two indigenous self-propelled guns in the 1970s.

**The Type 88 75mm AA gun was a further development of the Type 3 which went into service in 1928 and was their principal AA gun with field armies.**

A US Marine poses by a Navy Model 10 76mm AA gun, captured on one of the Pacific islands.

Right: At drill with the 75mm Type 88 AA gun.

## JAPANESE ARTILLERY EQUIPMENT

| Equipment | Date | Calibre (mm) | Barrel length (cals) | Breech mech. | Elevation max (deg) | Traverse (deg) | Weight in action (kg) | Shell weight (kg) | Muzzle velocity (m/sec) | Range, max (metres) |
|---|---|---|---|---|---|---|---|---|---|---|
| **Anti-tank artillery** | | | | | | | | | | |
| 37mm Mod 94 | 1934 | 37 | 45 | HSB/SA | 27 | 60 | 370 | 0.7 | 700 | 4600 |
| 37mm Mod Ra-97 | 1937 | 37 | 42 | HSB/SA | 16 | 90 | 450 | 0.9 | 800 | 4000 |
| 47mm Mod 01 | 1941 | 47 | 65 | HSB/SA | 19 | 60 | 755 | 1.03 | 830 | 7675 |
| | | | | | | | | | | |
| **Anti-aircraft artillery** | | | | | | | | | | |
| 40mm Mod 91* | 1931 | 40 | 39 | Auto | 85 | 360 | 890 | 0.8 | 610 | 3950 ceiling |
| 75mm Mod 88 | 1928 | 75 | 44 | HSB/SA | 85 | 360 | 2440 | 6.5 | 720 | 8850 ceiling |
| 75mm Type 4 | 1944 | 75 | 56 | HSB/SA | 85 | 360 | 3400 | 6.5 | 860 | 10050 ceiling |
| 80mm Taisho 10 | 1921 | 76.2 | 40 | VSB | 75 | 360 | 2400 | 6.1 | 680 | 7700 ceiling |
| 80mm Model 99 | 1939 | 80 | 45 | VSB/SA | 80 | 360 | 6575 | 9 | 800 | 9750 ceiling |
| 105mm Taisho 14 | 1925 | 105 | 40 | HSB/SA | 85 | 360 | 4480 | 15.8 | 700 | 10950 ceiling |
| 120mm Type 3 | 1943 | 120 | 65 | HSB | 90 | 360 | 22000 | 26 | 855 | 14650 ceiling |
| 15cm Prototype | 1944 | 150 | 60 | HSB | 85 | 360 | 54800 | 44.4 | 930 | 19000 ceiling |
| | | | | | | | | | | |
| **Mountain artillery** | | | | | | | | | | |
| 75mm Meiji 41 | 1908 | 75 | 19 | SQF | 40 | 7 | 544 | 6 | 435 | 7020 |
| 75mm Meiji 41 Improved | 1917 | 75 | 19 | SQF | 40 | 6 | 544 | 6 | 435 | 7100 |
| 75mm Model 94 | 1934 | 75 | 21 | HSB | 45 | 45 | 535 | 6.5 | 385 | 8300 |

Left: The 8cm High Angle Gun Type 3 first went into service in 1914. It was later improved to become the Navy Model 10.

Right: The 37mm anti-tank gun Type 94 was principally used as an infantry support weapon. This one is on Makin Island in 1943.

Inset: Loading a 28cm howitzer during the siege of Port Arthur in 1904.

| *Japanese Artillery Equipment (Cont)*<br>Equipment | Date | Calibre<br>(mm) | Barrel<br>Length<br>(cals) | Breech<br>mech. | Elevation<br>max<br>(deg) | Traverse<br>(deg) | Weight<br>in action<br>(kg) | Shell<br>weight<br>(kg) | Muzzle<br>velocity<br>(m/sec) | Range max<br>(metres) |
|---|---|---|---|---|---|---|---|---|---|---|
| **Field artillery** | | | | | | | | | | |
| 37mm Infantry Gun Taisho 11 | 1922 | 37 | 25 | VSB/SA | 14 | 33 | 94 | 0.6 | 450 | 5000 |
| 70mm Infantry Howitzer M92 | 1932 | 70 | 9 | SQF | 50 | 45 | 212 | 3.8 | 200 | 2800 |
| 75mm Field Gun Meiji 41 | 1908 | 75 | 30 | SQF | 16 | 12 | 980 | 5.7 | 510 | 10900 |
| 75mm Field Gun Meiji 38 | 1905 | 75 | 30 | SQF | 16 | 7 | 945 | 5.9 | 510 | 8250 |
| 75mm Fd Gun Meiji 38<br>Improved | 1917 | 75 | 30 | HSB | 43 | 7 | 1135 | 6.6 | 600 | 11960 |
| 75mm Fd Gun Model 90 | 1930 | 75 | 44 | HSB | 43 | 43 | 1400 | 6.5 | 700 | 14950 |
| 75mm Fd Gun Model 95 | 1935 | 75 | 31 | HSB | 43 | 50 | 1105 | 6.5 | 500 | 10950 |
| 105mm Gun Meiji 38 | 1905 | 105 | 32 | SQF | 35 | 6 | 3610 | 18 | 540 | 10050 |
| 105mm Gun Taisho 14 | 1925 | 105 | 34 | SQF | 43 | 30 | 3110 | 15.8 | 620 | 15000 |
| 105mm Gun Mod 92 | 1932 | 105 | 45 | SQF | 45 | 36 | 3720 | 15.8 | 760 | 18250 |
| 105mm How Mod 91 | 1931 | 105 | 24 | SQF | 45 | 40 | 1495 | 15.8 | 545 | 10765 |
| | | | | | | | | | | |
| **Heavy artillery** | | | | | | | | | | |
| 120mm Fd How Meiji 38 | 1905 | 120 | 12 | SQF | 43 | 64 | 1255 | 20 | 275 | 5760 |
| 15cm Gun Model 89 | 1929 | 150 | | IS | 43 | 40 | 10400 | 45.8 | 685 | 19900 |
| 15cm How Meiji 38 | 1905 | 150 | 12 | SQF | 43 | 3 | 2085 | 35.9 | 275 | 5900 |
| 15cm How Taisho 4 | 1915 | 150 | 22 | VSB | 65 | 6 | 2800 | 41.7 | 410 | 9550 |
| 15cm How Model 96 | 1936 | 150 | 22 | SQF | 65 | 30 | 4135 | 31.2 | 805 | 11850 |
| 24cm How Meiji 45 | 1912 | 240 | | | | | | 181 | 365 | 10335 |
| 41cm Siege Howitzer | | 410 | 30 | IS | 45 | 360 | 81280 | 997 | 535 | 19380 |

**Left: The 120mm Naval AA gun Model 10 was frequently removed from ships and installed as a combination coast defence and anti-aircraft weapon.**

**JOINTED GUN** A gun made in two pieces, muzzle section and breech section, for transport and joined together by some form of junction nut for firing. Commonly called the 'screw gun' it was invariably found with mule-pack artillery to keep loads at an acceptable weight for each mule. Advanced designs incorporated some form of safety interlock that prevented the gun being loaded and fired if the junction of the two parts had been incorrectly joined. Early designs did not have this safety device which occasionally produced some innocent entertainment among the gunners.

**JUMP** The vertical movement of the gun barrel due to the shock of firing, and which causes the axis of the bore to alter from the angle at which the gun has been laid. Jump is a constant value for a given gun, charge and shell combination and can be compensated for in the design of sights or in the firing tables.

*Japanese Artillery Equipment (Cont)*

| Equipment | Date | Calibre (mm) | Barrel Length (cals) | Breech mech. | Elevation max (deg) | Traverse (deg) | Weight in action (kg) | Shell weight (kg) | Muzzle velocity (m/sec) | Range max (metres) |
|---|---|---|---|---|---|---|---|---|---|---|
| **Coast artillery** | | | | | | | | | | |
| 10cm Gun Taisho 7 | 1918 | 105 | 45 | HSB/SA | 20 | 360 | | | 700 | 10050 |
| 10cm AA/CD Model 98 | 1938 | 100 | 65 | HSB/SA | 90 | 360 | 3900 | 13 | 1000 | 18650 |
| 127mm AA/CD Taisho 3 | 1914 | 127 | 50 | SQF | 77 | 360 | | 23 | 910 | 18450 |
| 15cm Gun Type 96 | 1936 | 149 | 50 | SQF | 50 | 360 | | 50.3 | 885 | 22850 |
| 15cm Gun Meiji 45 | 1912 | 149 | 50 | SQF | 43 | 360 | | 55.7 | 885 | 23750 |
| 20cm Turret Gun | | 206 | 47 | IS | 30 | 360 | | 126 | 762 | 18300 |
| 25cm Turret Gun | | 254 | 47 | IS | 35 | 360 | | | 810 | 24700 |
| 30cm Howitzer Taisho 7 Long | 1918 | 305 | 24 | IS | 73 | 360 | 20015 | 498 | 490 | 15250 |
| 30cm Howitzer Taisho 7 Short | 1918 | 305 | 16 | IS | 73 | 360 | 14900 | 400 | 396 | 10975 |
| 30cm Turret Gun | 1928 | 305 | 50 | IS | 33 | 270 | | | 850 | 29450 |
| 40cm Turret Gun | 1932 | 410 | 45 | IS | 35 | 270 | | 1460 | 762 | 30000 |

**Notes:** * Vickers design — also in twin form

### JAPANESE SELF-PROPELLED ARTILLERY

| Equipment | Date | Calibre (mm) | Chassis | Elevation max (deg) | Traverse on mount (deg) | Weight in action (kg) | Range, max (metres) | Ammunition carried (rds) |
|---|---|---|---|---|---|---|---|---|
| Type 38 Ho-Ro | 1942 | 150 | T97 tank | 30 | 8 | 15000 | 5950 | 15 |
| Type 74 | 1974 | 105 | Special | | 360 | 16500 | 11500 | |
| Type 75 | 1975 | 155 | Special | 65 | 360 | 25300 | 19000 | 28 |

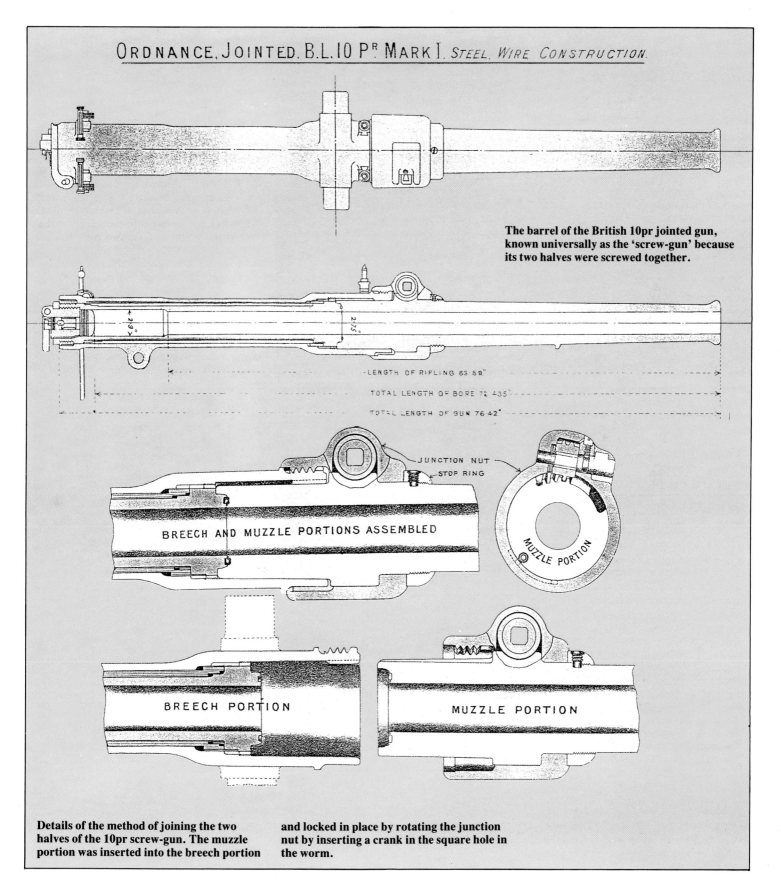

ORDNANCE, JOINTED. B.L. 10 Pʳ MARK I. *STEEL, WIRE CONSTRUCTION.*

The barrel of the British 10pr jointed gun, known universally as the 'screw-gun' because its two halves were screwed together.

LENGTH OF RIFLING 63.89"

TOTAL LENGTH OF BORE 72.435"

TOTAL LENGTH OF GUN 76.42"

JUNCTION NUT
STOP RING

BREECH AND MUZZLE PORTIONS ASSEMBLED

MUZZLE PORTION

BREECH PORTION

MUZZLE PORTION

Details of the method of joining the two halves of the 10pr screw-gun. The muzzle portion was inserted into the breech portion and locked in place by rotating the junction nut by inserting a crank in the square hole in the worm.

# K

**KARL** A German super-heavy self-propelled howitzer, developed by Rheinmetall-Borsig (qv) at the request of the German Army, who required a piece of heavy artillery that could be used where the lack of tracks made the use of railway guns impossible. This request was put forward in 1936, and it seems obvious that the German Army was looking east even then – there was no shortage of railway lines in any other direction.

The chassis was a fairly simple box structure carried on tracks supported by 11 roadwheels on each side. The wheels were carried on torsion bars connected to electrical gearing that allowed the bars to be rotated, so retracting the wheels and allowing the hull to be lowered to the ground for firing. A top carriage was free to move inside the hull, controlled by a hydro-pneumatic recoil system. The howitzer, of 60cm, was fitted into a ring cradle that had a separate recoil system connected to the breech ring of the weapon. Thus the howitzer had a dual recoil system between the barrel and the ground. The barrel could be elevated to 70° and the top carriage was capable of 8° of traverse on the mounting. The 60cm howitzer fired a 2,195kg (4,839lb) concrete-piercing shell to a range of 4,500 metres (4,921 yards) and a 1,575kg (3,472lb) high explosive shell to 6,675 metres (7,300 yards).

Six of these equipments were delivered to the Army between November 1940 and August 1941. After using them for some time the Army asked if it was possible to achieve a better range, and in May 1942 six 54cm barrels were ordered, to be interchangeable in the mountings with the 60cm barrels. Fitted with this barrel the equipment would fire a 1,250kg (2,756lb) shell to a range of 10.5 kilometres (6.5 miles).

The 54cm barrels were delivered by August 1943 and thereafter appear to have been interchanged with the 60cm quite randomly. Fitted with either barrel, the entire equipment weighed 124 tonnes (122 tons), was 11.3 metres (37 feet) long and stood 4.8 metres (15.7 feet) high. The vehicle could move for short distances under its own power, but for longer moves was carried on special rail or road transporters.

The weapons were taken to the Eastern Front and were used at Sevastopol, Lvov, Brest-Litovsk and other siege engagements. The six were named Adam, Eve, Thor, Odin, Loki and Ziu; the class name 'Karl', derives from General Karl Becker of the German Artillery, who instigated the project. Two were captured, in damaged condition, by US forces in Bavaria. Nothing is known to remain of these equipments.

**KELLY MOUNT** A form of ground platform based on the 'Panama Mount' (qv) and developed by the US Army to mount the 155mm Gun M1 in the coast defence role. It consisted of a central platform upon which the gun carriage rested, and a circular or part-circular track upon which the trail ends could ride, so that the gun could be slewed through large arcs of traverse quite quickly. Officially the 'Mount T6E1', it took its common name from Colonel P.E. Kelly of the Harbor Defenses of San Francisco, who designed it in 1943.

**A Kelly mount with 155mm gun deployed in the Pacific Theater during World War Two.**

Karl, with its tracked ammunition carrier
lifting a 1,250kg (2,600lb) shell.

**KROMUSKIT** American recoilless guns developed during the Second World War. They took this name from the designers, William J. Kroger and C. Walton Musser, both employees of Frankford Arsenal. Two types were developed, a 57mm and a 75mm, both using perforated brass cartridge cases and projectiles with pre-rifled rotating bands. The Dominion Engineering Works of Canada built 1,238 57mm guns as no production facilities were available in the USA, and 951 75mm example were built by the Miller Printing Machine Co. of Pittsburg, USA. A total of 100 guns was sent to Europe in March 1945, but it is believed that none was in use before the end of the war in Germany. A larger number was sent to the Pacific and used on Okinawa to good effect. The guns remained in US service until the middle 1950s, and copies are still widely used in various countries.

**KRUPP** German steel-making family which became famous for the manufacture of guns.

Friedrich Krupp (1787-1826) was a native of Essen and employed in an ironworks there. He became interested in the production of cast steel and with a partner, Friedrich Nicolai, began manufacturing it by a secret process in 1812. The business was not particularly successful but was continued by his widow and his son Alfried (1812-87). In about 1850 Krupp began making seamless tyres for German railway rolling stock and his business prospered. He exhibited blocks of steel at international exhibitions and began experimenting with solid steel as a gun barrel material. His first notable orders for guns were from the Egyptian Army in 1856, the Belgian Army in 1861 and the Russian Army in 1863, some of the latter weapons being of 24cm calibre. In the 1860s he developed the sliding block breech mechanism, and in 1865 he adopted the Armstrong system of building up guns by shrinking steel hoops around the barrel. His guns were used by the Prussian Army during the war of 1870, and in 1872-5 he completely re-armed the entire German Army and Navy and went on to supply guns to innumerable foreign countries. At the Columbian Exposition of 1893 in Chicago, Krupp exhibited a 24cm coast defence howitzer that was the progenitor of 'Big Bertha' (qv). Alfried was succeeded by his son Friedrich Alfried (1854-1902), who left only a daughter, Bertha. She married Gustav von Bohlen und Halbach in 1906, and upon marrying he took the style of Krupp von Bohlen und Halbach and became head of the firm.

The company expanded to take in a number of lesser steelmaking concerns, and also built shipyards in Kiel. After the First World War the company turned to the manufacture of railway rolling stock, ironmongery, trucks, dynamos, cameras and many other items, but gun design staff was farmed out to other companies in Europe where it could be employed without contravening the requirements of the Versailles Treaty. By 1933, when Hitler acceded to power, Krupp was back in the gunmaking businesss and manufactured a wide range of artillery for the new German Army and

Navy and also for export in the 1930s. The company was seized by the British Military Government after the Second World War with the declared intention of dismantling the firm and preventing it from ever being reassembled. However, various protracted political manoeuvres followed and the firm

**A Krupp 76mm 'balloon gun' of 1909.**

emerged as whole as ever. It is currently part of a major industrial consortium and manufactures armoured vehicles, but not guns.

Left: Loading an American 106mm recoilless gun, showing the perforated cartridge case typical of the Kromuskit design.

Above right: Not all Krupp designs were successful; this 45mm anti-tank gun was turned down by the German Army in 1937.

Right: Assembling 21cm heavy howitzers in the Krupp factory at Essen, 1942.

**KRUPP NON-RECOIL MOUNTING** A type of mounting devised by Krupp for installation in armoured cupolas used in land forts. It consisted of a conventional gun which had the exterior of the muzzle shaped into a ball; this was mounted into a socket in the armoured cupola so that the gun could elevate using the ball as a pivot: in essence, a muzzle-pivoting (qv) mounting. However, the significant feature of this mounting was that the force of recoil was passed via the ball and socket mount into the structure of the armoured cupola, and from that, via its foundations, into the ground. Thus the gun barrel did not move in recoil, and could be rapidly loaded. The system was only feasible because of the enormous mass of the armoured cupola and its underpinnings.

A Krupp non-recoil mounting showing how the gun muzzle was anchored in the armour plate. The gunner's position was more popular with the designer than with the men who had to fire the weapon.

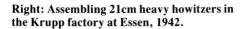

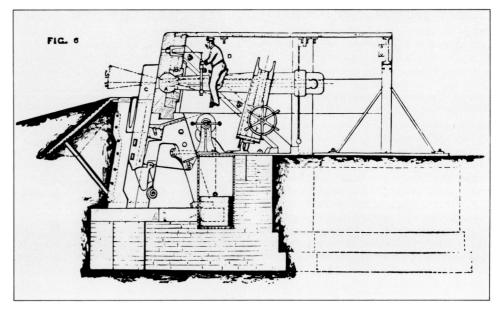

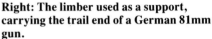

Right: The limber used as a support, carrying the trail end of a German 81mm gun.

Left: A typical ammunition limber of the horse-drawn period, open to show the ammunition which is fitted with clips and tape handles to facilitate withdrawing the fired rounds.

# L

**LANCASTER** A system of 'rifled' ordnance put forward in the 1850s in Britain. The Lancaster gun had a smooth bore of oval section which was then twisted. The projectile had a similar oval cross-section and was muzzle-loaded. On firing, the mechanical fit between bore and shot made the shot spin as it followed the twisted oval of the bore. A number of Lancaster guns were manufactured and sent to Sevastopol during the Crimean War but the guns were not a success. The shot had a tendency to jam in the bore and damage the interior surface, and the already poor accuracy became even worse as a result. They were withdrawn from service and were left in the Crimea when the war ended.

**LANGER EMIL** A German 38cm ex-naval gun on a special barbette mounting that the German Army emplaced on the Belgian coast in 1915 and with which they shelled Dunkerque at a range of about 40 kilometres (25 miles). When the bombardment began, it was the first long-range artillery attack ever attempted and the Allies were convinced that the explosions were due to Zeppelin bombs. Examination of the projectile fragments eventually convinced them that it was actually a gun doing the damage, but in spite of aerial reconnaissance and bombing attacks on likely spots, Langer Emil was never found and remained in business until the end of the war.

**LAYING** Artillery term for pointing the gun at the target, either in direct or indirect fire. In British service the man who operates the sight is known as the 'layer'; in American service, as the 'gunner'.

**LEAD** The angular offset between sight and gun barrel required to hit a moving target in direct fire.

**LEVELLING GEAR** Apparatus used on mobile gun mountings, particularly anti-aircraft guns, to ensure that the platform of the gun is perfectly level and that the gun when it traverses does not change its elevation or vice versa. Usually it takes the form of screw-jacks mounted at the end of the outriggers (qv) or at the corners of the platform, though some equipments have had a sub-platform that could be levelled, by the same means, independently of the main platform. Modern technology has substituted electronic sensing systems that automatically apply corrections to the sights, so that the need for precise levelling is no longer so important.

**LIMBER** A wheeled carriage that can be attached to the trail end of a gun to support it and so convert the two-wheeled gun into a four-wheeled trailer for transport. In light artillery it was often used as a means of carrying a supply of ammunition for immediate use; with heavier weapons it is usually little more than a two-wheeled axle with a drawbar and some form of connection for the gun trail. In American service it is known as a 'Caisson'.

**Left: The limber as an ammunition trailer, seen here with an American 75mm gun M1917A1.**

**Above: A German 128mm K44 anti-tank gun, showing the levelling jacks at the ends of the outriggers.**

Below: The 914mm Little David mortar emplaced in a pit.

The barrel of Little David being towed.

**LINER** A tube, containing the chamber and rifling, inserted into a gun. In broad terms a liner resembles an 'Inner "A" Tube' but it differs in being a mechanical fit inside the 'A' tube (qv) and is not initially affected in the building-up of the gun, nor does it contribute to the longitudinal strength of the gun. Liners can be used for repairing a worn gun by removing the original liner or Inner 'A' tube and replacing with a new liner; to form the bore of a new gun in such a manner that repair can subsequently be easily carried out by removal and replacement of the liner; or to change the calibre of an existing gun. As an example of the latter, the British 3.7in Mark 6 gun was a 4.5in AA gun with its 4.5in liner removed and replaced by a new one of 3.7in calibre. This substitution enabled a longer barrel to be employed, and made for a far stronger gun due to the greater thickness of the original 4.5in gun body; a more powerful cartridge could thus be employed.

**LITTLE DAVID** Little David was a 914mm muzzle-loading mortar developed by the US Army in 1944-5. It was originally developed as the 'Bomb Testing Device T1', a method of projecting aerial bombs for test purposes. When testing piercing bombs it is necessary to deliver them with some accuracy on to a suitable hard target, and aircraft dropping was rarely accurate enough. Old howitzers were therefore adapted for the purpose, firing the bombs out with a reduced charge sufficient to give them the correct striking velocity. It became necessary to develop a special weapon as bombs grew bigger, and the US produced a 36in mortar. The idea arose in March 1944 of converting this test device into a service weapon, the incentive being the probability of encountering heavy fortifications during the invasion of Japan and the nearby islands. By April 1945 the equipment was built and tested and discussion began about its possible employment in Operation 'Olympic', the projected invasion of Japan scheduled for November 1945. The Coast Artillery Board agreed to give the weapon its service acceptance test, but the Army Ground Forces complained of poor accuracy in the initial testing, and while this was being thrashed out the war ended. Further work was suspended and the mortar was sent to Aberdeen Proving Ground, where it is still on display.

The mortar was a rifled muzzle-loader supported in a ring cradle and with an elevating arc attached to the breech end. The mounting was simply a large steel box with trunnion bearings which was sunk into a suitable pit. Barrel and mounting were separately transported, the barrel being semi-trailed on an eight-wheel dolly behind a tank transporter tractor, the mounting being fitted with wheels for towing behind another tractor. The shell was a peculiarly-shaped projectile with hemispherical base and a long pointed nose. The charge of 160kg (353lb) of smokeless powder was loaded into the muzzle; the shell was then loaded by means of a crane into the muzzle, the pre-engraved driving band being engaged with the rifling. The barrel was then elevated so that the shell slid down

**Four men carry a 195kg (410lb) shell to a 240mm howitzer on its loading tray.**

**The complicated firing lock on a British 6in coast defence gun, designed to permit firing by either percussion or electrical means.**

to the shot seating ahead of the chamber and the mortar was then fired by a percussion primer. The shell contained 725kg (1,598lb) of high explosive and had a range of 8,685 metres (9,498 yards).

**LOADING TRAY** A curved tray for ammunition to facilitate loading into the breech of the gun. It is used in equipments where the projectile or complete round is too heavy to be easily hand-loaded and thus needs to be lifted by two men, or where it is advantageous to have the round or components of it positioned for ramming by a mechanical rammer.

**LOCK** An abbreviated form of 'firing lock'; the firing mechanism used with bag charge guns, which is assembled to the rear end of the breechblock. It may fire the gun by percussion, electric current, or be a combination lock which allows the use of both methods alternatively.

**LONGHAND** British medium anti-aircraft gun, the last 'cannon solution' to be developed and approved for service against high-altitude aircraft. It consisted of a wartime 3.7in Mark 6 medium gun fitted with a 12-round rapid-loading conveyor system which gave it a cyclic rate of fire of about 80 rounds per minute. Approved for service in February 1957, production had not begun when the decision was taken in 1958 to adopt guided missiles for heavy air defence, and Longhand was retired.

**LONG TOM** A name frequently applied to various long-barrelled guns, and used by the Boers in South Africa to describe a number of 15cm Krupp guns they employed against Ladysmith and Mafeking. More recently it was the 'official nickname' for the American 155mm Gun M1 during and after the Second World War.

**LOOSE LINER** A rifled barrel that fits inside the remainder of the gun, taking the form of an 'A' Tube (qv) except that it does not contribute to the mechanical strength of the gun structure. It can be removed relatively easily for replacement in the field when the rifling is worn.

**Longhand, or the 3.7in gun X4, the last medium-calibre British anti-aircraft gun to be developed for the British Army.**

**LUELLEN-DAWSON** An American company that propounded a scheme for portable coast defence artillery in the 1920s. The idea involved preparing reinforced concrete bases at suitable points round the coastline and linking them to the nearest railway line. Special trains of guns would be held at strategic rail junctions and, in the event of a threat materializing, would be dispatched to the threatened area to disperse the guns around the prepared bases. The idea was enthusiastically embraced for a short time and then abandoned.

**LYMAN AND HASKELL** Two American inventors who appear to be the originators of the multiple-chamber gun concept. The Lyman and Haskell gun tested at Frankford Arsenal in 1870 was a 6in weapon, basically like an Armstrong gun but with only three rifling grooves and three ribs on the shell. In addition to the conventional chamber at the rear of the bore for the reception of the cartridge, there were four additional chambers, each with its own cartridge, connected to the barrel at intervals along its length. The basic cartridge was 1kg (2.2lb) of black powder, and a 3kg (6.6lb) charge was placed in each of the additional chambers. A wad behind the shell acted as a gas seal. The theory was that the basic charge would be fired in the normal way and would begin moving the shell up the barrel. As the charge passed along and uncovered the entrance to the first auxiliary chamber, so the flash would enter and fire the auxiliary charge, adding more gas and pressure and increasing the velocity of the moving shell. As the successive chambers were exposed, so their charges would fire until the shell had reached a very high velocity, and could achieve an extremely long range.

The gun produced a muzzle velocity of only 335m/sec (1,099ft/sec) when tested, which was far below what a conventional gun might have been expected to give, even though the pressures recorded in the barrel were as high as 38 tons to the square inch; a figure that would make even a present-day ordnance engineer think twice. Investigation showed where the fault lay: the flames from the initial charge went past the shell and ignited the auxiliary chambers before the shell reached them, so that it was fighting its way up the bore against all sorts of pressures. The US government rejected the idea and the Lyman and Haskell gun vanished into history.

The idea, though, refused to die, and reappeared at intervals. Its last appearance was in the German High Pressure Pump (qv) project in 1942-5. It seems to be due for a revival about now.

# M

**MEAN POINT OF IMPACT (MPI)** The arithmetical centre of the points of impact of a group of shots.

**MECHANICAL LOCK** A term that relates to the operation of a gun breech mechanism. The breech, when closed, is subjected to very high pressure when the cartridge explodes, and there must be some positive lock to prevent this pressure blowing the breech open. Usually the arrangement is very simple: the linkage from the operating lever passes 'over dead centre' before coming to rest in the closed position, so that any opening movement of the breechblock will tend to force the operating lever tighter against its stop. As the lever is operated to open, so the linkage passes through dead centre again before beginning the opening movement of the block.

**METEOR** Abbreviation for 'meteorology', which refers to the application of corrections to the calculated elevation and azimuth necessary to hit the target. These corrections are deduced from current meteorological data and will compensate for air temperature, charge temperature, crosswinds and barometric pressure, all of which affect the flight of the shell. The information is derived from radio-sonde equipment and passed to artillery units in the form of a 'Meteor Message' at regular intervals. This is then used to deduce corrections, either by tables, by graphical means or, in modern artillery systems, by computer.

**MIKE TARGET** British Army abbreviation for a 'regimental target'; ie, one engaged by all the guns of a regiment. It was part of the system of fire control perfected during the Second World War which allowed one observer to bring down fire onto a specific target from an entire regiment. The system extended upwards to Uncle (divisional), 'Victor' (Corps) and 'Yankee' (Army) targets. The names come from the phonetic alphabet in use in the 1940s.

**MIL** A unit of angular measure used by artillery for aiming and survey purposes. One Mil is 1/6400th part of a circle; it is loosely derived from radian measure but the derivation has been adjusted so that one mil is the angle subtended by one unit of distance at 1,000 units range. Thus, two stakes 1 metre (3.28 feet) apart, planted 1,000 metres (3,280 feet) from an observer, will measure an angle of one mil between them. It is equivalent of 17.7778 minutes of arc.

**MISFIRE** A total failure of a gun to fire; usually attributed to an ammunition fault, but equally likely to be a mechanical fault in the gun. In the case of cased-charge guns such faults can include weakness of the striker spring, wear on the tip of the firing pin, or setting-back of the breechblock, all of which reduce the blow of the striker. The same faults can be found in the firing lock in bag-charge guns. In addition, fouling can block the vent, preventing the flash igniting the charge.

**MOBELWAGEN** German anti-aircraft self-propelled equipment, in use in 1944-5. It consisted of the chassis of a Panzer IV tank surmounted by a flat bed on which a 3.7cm FlaK43 gun was mounted, enabling it to traverse through 360°. During movement, folding sides enclosed the gun and crew and offered light armour protection, but when in action the sides were allowed to drop down and became a platform around the gun. The equipment was used for mobile protection of armoured columns and 240 were built. The name (which means 'furniture van') was bestowed by the troops because of the vehicle's high, square outline.

**MOBILE MOUNTING** A type of gun mounting that has wheels attached to move it from place to place but goes into action without its wheels. Commonly seen on medium and heavy anti-aircraft guns. The word 'mobile' is often used with some anti-aircraft guns to distinguish them from 'static' guns that were moved on special transporters and secured into prepared emplacements for use. In broad terms a mobile gun will usually be built with an eye to lightness, for transport and for less stress on the mounting, while static mountings can be made with heavier components; for example, the British 3.7in gun in mobile form used hydro-pneumatic balancing gear to balance the barrel weight, while the static version used a massive counter-weight over the breech. The latter solution was cheaper and easier to make, but much heavier and inappropriate to a mobile mounting.

**MONCRIEFF** Colonel Sir Alexander Moncrieff (1829-1906) was the inventor of the first practical disappearing gun carriage. The son of a captain in the Madras Army, Moncrieff became a lieutenant in the Forfarshire Militia Artillery in April 1855, and forthwith obtained leave to go to the Crimea, where he was present at the siege of Sevastopol. Promoted to captain in September 1857, he transferred to the Edinburgh Militia Artillery in November 1863. He was promoted major in March 1872, and Colonel of the 3rd Artillery Brigade in 1878.

He was present in the Crimea during the bombardment of the Mamelon on 6 June 1855 and, struck by the vulnerability of the Russian guns, which were being hit through the gaps in their embrasures, he began contemplating some method of allowing guns to rise and fall behind a parapet between shots. He perfected his design, which used two curved arms to support the gun, balanced by a counter-weight, during his service with the Edinburgh Artillery. The idea was initially rejected but later accepted by the Royal Artillery. Moncrieff spent from 1867 to 1875 at the Royal Arsenal, Woolwich, superintending the development and manufacture of his carriages. Seeing that there were limitations to his design, in 1869 he first submitted ideas for a hydro-pneumatic mounting. This design relied on an hydraulic ram to support the gun in the 'up' position and control the recoil, and was eventually adopted in many countries. He received an award of £10,000 for his invention of the original disappearing carriage. Moncrieff became a Fellow of the Royal Society in 1871, and a Knight Commander of the Bath in 1890.

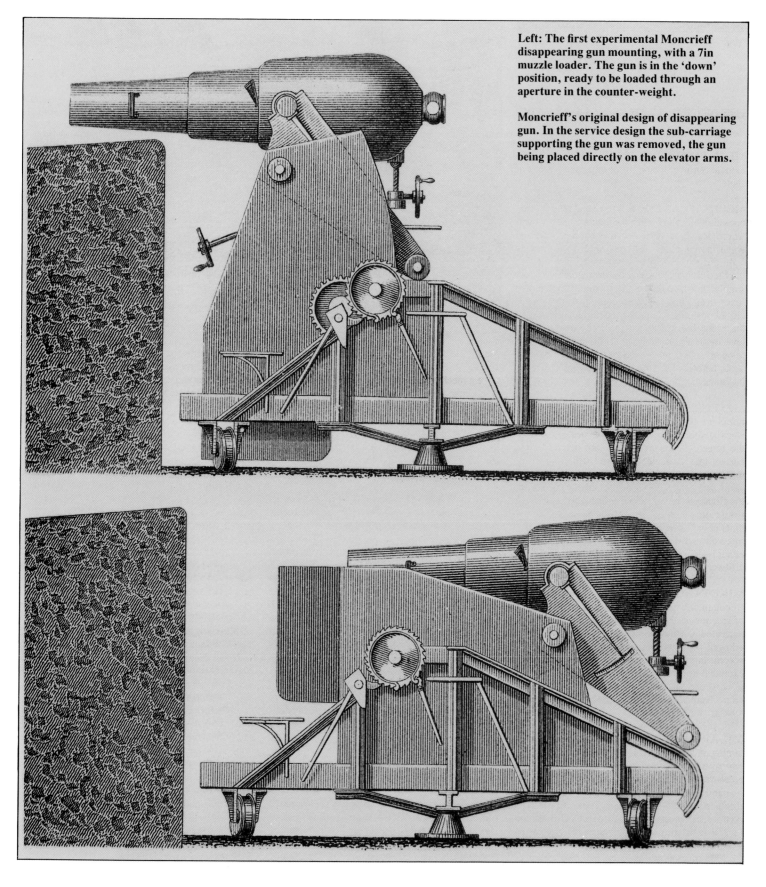

**Left: The first experimental Moncrieff disappearing gun mounting, with a 7in muzzle loader. The gun is in the 'down' position, ready to be loaded through an aperture in the counter-weight.**

**Moncrieff's original design of disappearing gun. In the service design the sub-carriage supporting the gun was removed, the gun being placed directly on the elevator arms.**

**MONOBLOC** A gun barrel made from one piece of metal. Used originally to distinguish such guns from those of built-up construction, today its use is superfluous as almost all modern guns are monobloc, usually auto-frettaged (qv) to increase their strength.

**MORTAR** Strictly, any piece of ordnance designed to fire only at angles of elevation greater than 45°. (Compare, in this respect, with 'howitzer' and 'gun'.) Today's mortars are invariably short-barrelled infantry weapons that do not come under the classification of 'artillery' and will not be further considered, even though some large-calibre Soviet mortars have a performance that almost equals that of field howitzers.

In the artillery field, the term refers to specialized howitzers that were used in the past for siege bombardments and for coast defence. Their advantage was that they delivered their shells in a steep downward trajectory, which attacked the enemy from above rather than battered at his side defences

which, in the case of siege weapons, would be immense masonry and earth ramparts or, in the case of coast defence weapons, would be the thick armoured side of a battleship. Attack from the top meant that the mortar attacked the roof of the fort or the deck of the ship, either of which would be a great deal thinner than the side defences. In the case of forts, the mortar was negated by developing overhead protection and by sinking the fort into the ground; the solution for ships was never really found, as it was impossible to develop sufficient thickness on the upper surfaces and retain the vessel's stability. As a result, many countries used mortars in coast defence until the 1950s.

In spite of the common present-day conception of a mortar as a muzzle-loaded weapon, artillery mortars were no more than short-barrelled breech-loading howitzers. In order to obtain shells of sufficient weight and power to do what was needed, they were generally of calibres between 30 and 45cm firing shells weighing up to 700-800kg (1,543-1,764lb).

**Above: American 12in M1890 coast defence mortars in their sunken emplacement.**

**Above right: German mountain troops with a 75mm Gebirgskanone in the Caucasus mountains.**

**MOUNTAIN GUN** A gun or howitzer specifically designed for operation in mountainous terrain, which means it can be dismantled into its principal components and carried by mule or man-pack in the most difficult conditions. Lightness is the most important property, coupled with shell-power and a reasonable range. High elevation is also desirable, since mountainous country usually means firing over the top of ridges and peaks to reach the enemy.

**Right: A selection of muzzle brakes, illustrating the various approaches to the problem of reducing recoil.**

**MUSHROOM HEAD** Part of the breech mechanism of a bag-charge gun. The mushroom head, so-called from its shape, consists of the head – which forms the face of the breech and closes the rear of the chamber – and its 'stalk' or spindle, which passes through the breechblock and is retained by a nut. The vent passes down the axis of the mushroom head and the firing lock is attached at the rear end. The obturating pad (qv) is trapped between the front face of the breechblock and the rear face of the mushroom head.

**MUZZLE BRAKE** A device fitted to the muzzle of a gun to divert some of the gas that emerges in the wake of the projectile, turning it to develop a forward thrust and so counteract some of the force of recoil that is moving the gun backwards. The design of a muzzle brake has to be a compromise between efficiency and practicality. By turning a high proportion of the emerging gas through 180° and directing it backwards, it is possible to reduce the recoil by a very large amount, but the gas so directed backwards would make the gunners' task untenable. The gas is therefore usually directed slightly backwards of the muzzle, and in an amount sufficient to make a worthwhile reduction in recoil but not sufficient to incommode the gun detachment. Moreover, much of the brake's effect is achieved by the force of the emerging gas acting upon the baffles, which direct the gas sideways; as the gas strikes these baffles it attempts to thrust the gun forward even as the gas is being redirected. A good muzzle brake can be expected to be of some 25 to 35 percent efficiency – ie, it will reduce the recoil force by that percentage. Some ornate designs have improved considerably on this, but usually at the expense of difficulty of manufacture; certainly one highly efficient design, in the writer's knowledge, was spurned by a number of makers who claimed it was impossible to manufacture it economically.

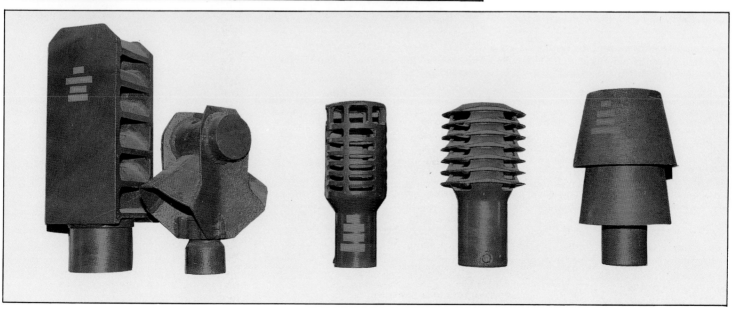

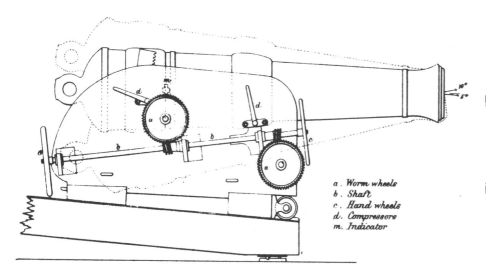

a . Worm wheels
b . Shaft
c . Hand wheels
d . Compressors
m . Indicator

**A diagram of Colonel Shaw's muzzle pivoting carriage of 1863. Turning the handwheel 'c' rotates the shaft 'b' and the worm wheels 'a' which operate spur gears on two curved racks (one of which is visible) to lift the rear of the gun.**

**MUZZLE ENERGY** The amount of kinetic energy contained in the projectile at the instant it leaves the muzzle. Normally calculated in foot-tons, it is based upon the formula $\frac{1}{2}MV^2$ and is arrived at in practice by multiplying the projectile weight by the square of the muzzle velocity and dividing the result by a factor that takes into account gravity and the varying units of measure used.

**MUZZLE PIVOTING CARRIAGE** A type of gun carriage developed in the period 1865-85 to arm forts. The object was to have a gun behind an armour-plate shield with a port cut in it to permit the gun to fire, but to keep the dimensions of the port no bigger than was necessary to allow the muzzle to poke through. In the first armoured forts, the port had to be large enough to allow the gun some degree of traverse from side to side, and this meant that there was the danger of shot or fragments coming through the port to damage the gun or injure the gunners. The Muzzle Pivoting Carriage (sometimes called the 'Small port carriage') was arranged so that as the gun was elevated and depressed it pivoted round an imaginary axis through the muzzle, minimizing the area of the port to no larger than that required to allow the muzzle through. The elevation mechanism usually consisted of twin sets of arcs that guided the movement, and an hydraulic jack to lift the gun bodily. Once the mechanism was

at the correct elevation, screw-jacks were rapidly run up to support the weight while the gun fired.

Muzzle-pivoting carriages were used with rifled muzzle-loading guns at some British armoured forts, notably Fort Cunningham, Bermuda, and Breakwater Fort, Plymouth, and similar designs were developed on the continent by Krupp and Schneider to mount breech-loaded guns in armoured cupolas in land forts.

**MUZZLE SWELL** The gradual increase of the external contour of the gun barrel close to the muzzle. It was adopted in the 1890s with the development of quick-firing guns with shields, as it was thought that the muzzle ought to be strengthened in order to withstand the impact of enemy bullets, and as it also added strength at the muzzle to withstand the higher pressures obtained by the use of smokeless powders. With the adoption of autofrettaged barrels, which gave more strength, the idea was gradually abandoned.

**MUZZLE VELOCITY** The speed of the projectile as it leaves the muzzle of the gun. It is normally quoted as a measure of the gun's power and given either in feet or metres per second.

**The Nordenfelt breech on a French 75mm Mle 1897 gun.**

**N**

**NAPOLEON** Common name given to the American Light 12pr Field Gun, Model of 1857. It was a smoothbore muzzle-loading field piece that had been originally designed for the French Army by Napoleon III and intended to rationalize the field artillery by replacing some lighter and heavier guns by a universal model. It was lighter than previous 12 pounders, having been designed to the specification that the gun should weigh 100lb for every one pound of shot weight. It was extensively used throughout the American Civil War and remained in general service for several years afterwards until replaced by modern rifled weapons.

**NON-RIGIDITY** A term used in connection with firing guns against ground targets at considerably different heights. A shell fired against a target that is on the same horizontal plane will describe a particular trajectory. If the target is raised or lowered several hundred metres above or below the gun and the gun elevated or depressed to hit it, the shape of the trajectory will not remain the same – ie, it will be 'non-rigid'. It will therefore be necessary to make a correction to the measured horizontal range between gun and target in order to compensate for the change in trajectory; in the case where the target is higher, the range will have to be increased, and reduced if the target is lower. This correction is referred to as a correction for non-rigidty of the trajectory.

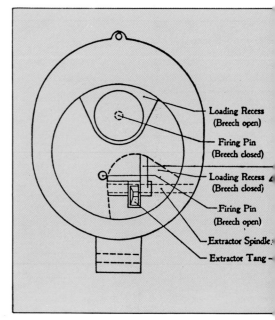

Loading Recess
(Breech open)

Firing Pin
(Breech closed)

Loading Recess
(Breech closed)

Firing Pin
(Breech open)

Extractor Spindle

Extractor Tang

**NORDENFELT** Torsten Nordenfelt (1842-1920) was a Swedish banker who was approached by an inventor, Helge Palmkranz, in about 1879 for financial backing for a machine gun design. Palmkranz got his backing on the understanding that the gun would be named 'Nordenfelt', and thereafter faded from the scene. Nordenfelt applied his business acumen, promoted the gun successfully, and founded the Nordenfelt Gun and Ammunition Company with factories in Sweden, Britain and Spain. The company subsequently developed a light 6pr gun for naval and coast defence use against torpedo-boats, and various other patents applicable to artillery and weapons.

**NORDENFELT BREECH** A type of breech mechanism for cased-charge guns, sometimes called the 'eccentric screw' system. It consists of a screw that is considerably greater in diameter than the gun, mounted with its axis away from the axis of the bore. There is a cut-out in the periphery of the screw which, when the screw is turned to the 'open' position, aligns with the chamber of the gun and permits the round to be loaded. When the screw is turned to 'closed', the cutaway is rotated away from the chamber and replaced by a solid part of the screw. As the screw rotates, so the pitch of its thread forces the screw towards the chamber, so supporting the cartridge case and bringing the firing mechanism into the correct position; conversely, as the screw opens, the pitch withdraws it slightly, allowing the extractors room to function and eject the empty case. Used in a number of Nordenfelt designs, its most famous and long-lasting application was in the French 75mm field gun; see 'French 75'.

**A diagram showing the various elements of a Nordenfelt breech mechanism as used on the French 75mm Mle 1897 gun.**

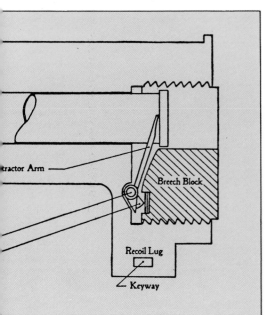

**OBSERVED FIRE** Indirect artillery fire controlled by eye from an observation post. The observer indicates the target by using a grid reference and then observes the fall of shot, deduces corrections, and orders these to the guns until such time as a target round is obtained, after which he may order fire for effect.

**OBTURATION** A technical term meaning the sealing of the gun breech against the unwanted escape of gas. Note the word 'unwanted'; this covers the case of a recoilless gun, where a proportion of the gas is required to escape to the rear.

Obturation is today obtained by one of three methods:
1, By means of a metal cartridge case enclosing the charge. The metal of the case expands under the pressure of the exploding charge, presses tightly against the walls of the chamber and thus effects the necessary seal.
2, By means of an obturator pad (qv) inside the breech screw mechanism. This system is used with guns having their cartridges in a cloth bag or-in a combustible case, where there is no metal case to make the seal.
3, By means of an expanding metal ring in the face of the breechblock which presses against the rear of the chamber. The Broadwell Ring (qv) was an early example of this system, and its modern counterpart is currently used in one or two tank guns that use bag charges but need to have a sliding blockbreech because of space limitations inside the tank turret.

**OBTURATOR PAD** A soft, doughnut-like ring that fits behind the mushroom head (qv) and in front of the breechblock on a bag-charge gun. In its earlier form it consisted of asbestos and rapeseed-oil confined within a brass wire mesh casing, but in modern weapons it is made of Neoprene or some similar heat-resistant plastic material. The pad is shaped to mate with a prepared coned surface in the mouth of the chamber. With the breech closed, the pad makes firm contact all round. When the cartridge is fired and the chamber pressure builds up, the mushroom head is pressed backwards; this squeezes the pad outwards to make a seal that is actually being applied to the chamber wall at a higher pressure than the gas trying to escape, so that a failure of the pad to seal is almost unknown.

The obturator pads can be classified as 'slow cone' or 'steep cone'. Early pads were slow cone, meaning that the angle of the pad seat was less than 10° and it was necessary to withdraw the breech screw and pad axially before the opening swing

**Two obturator pads. Top, an early design using canvas and wire. Bottom, a design dating from about 1914, using asbestos and wire mesh tightly compressed.**

could begin. Steep cone pads have a seating angle of about 26° and thus the breech screw can begin its opening swing immediately without the need for axial withdrawal. One of the few weapons using a slow cone obturation today is the American 155mm Howitzer M114.

**OSTWIND** A German self-propelled anti-aircraft gun consisting of a 3.7cm FlaK43 gun in an hexagonal open-toppped rotating turret on a modified Panzer IV tank chassis. It was developed to replace 'Wirbelwind' (qv) with an equipment having the more effective FlaK43 gun. Production commenced in December 1944 and some 40 were built before the war ended.

**OUTRIGGERS** These are arms attached to the four projections of a cruciform (qv) mounting and hinged so that they can be folded up for transport or down for supporting the gun in action. The outriggers make the effective base of the gun much wider than would be practical with a solid base, and thus are usually used where it is desired to allow the gun to traverse through 360° and fire at any point–as, for example, an anti-aircraft gun. The outrigger is usually locked when down, to provide a firm base, but at least one gun (the Skoda-designed 105mm FH43 intended for the German Army) had an hydraulic self-levelling apparatus that permitted the outriggers to lie loosely at whatever level the ground allowed but lock solid when subjected to the firing shock.

# P

**PACK HOWITZER** A short-barrelled weapon designed to be easily and quickly dismantled into its major components and carried on mule-back by the use of pack-saddles. The break-up varies according to the designer's ideas, but is generally as follows: the barrel, which may be divided into two parts; the breech ring and mechanism; the recoil mechanism and cradle; the trail which, again, may be in two parts (the axle and wheels). Pack howitzers may also, of course, be transported complete, by towing, or when dismantled they may be transported in other ways; eg, dropped by parachute.

**An Indian Army gunner and pack mule with the front half of the barrel of a jointed 10pr mountain gun.**

**PALLISER** Sir William Palliser (1830-82) was commissioned into the Rifle Brigade in 1855, transferred to the 18th Hussars in 1858, and retired with the rank of major in 1871. As an undergraduate at Cambridge he had become interested in guns and ammunition, and in 1862 he patented a method of lining smoothbore guns with a wrought-iron rifled liner, so allowing obsolete smoothbores to be converted into rifled muzzle-loaders. The gun was bored out to remove any surface imperfections, the liner inserted, and it was then expanded to a tight fit by simply firing a heavy charge of powder inside it. The 'RML Converted' guns were a cheap and effective method of providing large numbers of medium-calibre RML guns for short-range coast defence and were adopted in large numbers in Britain in the 1870s. Palliser was awarded £7,500 for the use of his patent. Similar conversions were later 'invented' by others in other countries.

Early in 1863 Palliser developed a method of casting iron shells with their noses in a cold iron mould, the remainder of the shell being in a sand mould, which gave the nose of the shell exceptional hardness. This produced a projectile with excellent armour-piercing capability in the days when 'armour' merely meant a considerable thickness of wrought iron. For this he received an award of £15,000. The shell was adopted in Britain in 1867 and remained in use until 1909; it was also adopted by several other countries during that period.

Palliser was awarded the CB in 1868 and the KCB in 1873. After retirement he became a Member of Parliament

**PANAMA MOUNT** A form of prepared emplacement developed by the United States Army. It consisted of a central pivot block of concrete carrying a pivot pin, surrounded by a circular ring of concrete carrying a steel racer ring. The dimensions were drawn up to suit the carriage of the 155mm GPF M1917 gun; this gun could be run on to the central block and the pivot pin connected to the axle-tree. The opened trail legs then rode on the steel racer on the outer concrete ring, allowing the gun to be pivoted around to cover a very wide arc. These platforms were installed at various points on the coastline where an enemy attack might take place but where the chances of such an attack were insufficient to warrant the mounting of permanent artillery. Thus, if a threat developed, field artillery guns could be emplaced and employed in the coast artillery role. The system was first adopted in

**An American 155mm M1917 gun on a Panama mount.**

Panama, which led to the name; it was then used elsewhere, and some of the Panama mounts in the Philippine Islands were actually used during the Japanese invasion in 1941-2. See also 'Kelly Mount'.

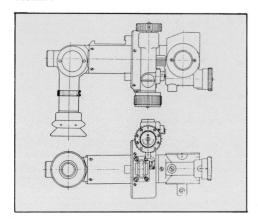

**Above: A typical panoramic sight; the portion above the setting knob rotates through 360° and its pointing is read in the window directly above the eyepiece.**

**Right: Using a panoramic sight on a 155mm FH70 howitzer.**

**PANORAMIC SIGHT** An optical gun sight that has a fixed eyepiece and a movable head so that the head can be rotated through 360° to look in any direction without the gunlayer having to move his eye. The movable head is provided with a suitable scale to measure the angular displacement of the head from a line parallel to the axis of the gun barrel. Originally called a 'goniometric sight', and sometimes called a 'dial sight', the panoramic sight allows the gun to be laid on an aiming point (qv) so it can engage in indirect fire (qv). It can also be set parallel with the gun barrel and used as a direct-fire sight in an emergency.

**PACQUETTE** A disc of steel that fitted round the fuze of the shell for the French 75mm M1897 gun and degraded the ballistic performance to give the shell a shorter range but a steeper angle of descent to the target. It was adopted during the 1914-18 war in order to overcome the shortage of field howitzers. Also called the 'Malandrin Disc' after its inventor.

**PARIS GUN** A German long-range gun used to bombard Paris from March to July 1918. Also known as the 'Kaiser Wilhelm Geschütz'. Developed by Krupp and designed by Professor Rausenberger, who had also designed Gamma (qv) and Big Bertha (qv), the Paris Gun was a 21cm gun constructed by inserting a 21cm (8½in) liner into a bored-out 38cm naval gun body. The liner actually projected some 13 yards beyond the muzzle of the parent gun and had to be supported by a girder bracing to prevent excessive droop. At the muzzle a smoothbore extension 6 metres (6½ yards) long was added, giving a total barrel length of 40 metres (43 yards) and a weight of about 145 tonnes. The mounting was a fairly simple box, pivoting about its front end and with wheels at the rear which ran around a prepared track. The gun was carried in a ring cradle trunnioned to the sides of this box and was able to recoil about 2 yards. The front support was bolted down to a massive concrete emplacement built in the Forest of Crépy, about 80 miles from Paris. The full details of the Paris Gun have never been revealed, but it is believed that three mountings and seven barrels were built. The first gun fired for three days and then closed down to have the barrel changed, since the rate of wear was atrocious – the forcing cone in front of the chamber moved several centimetres up to the bore with each shot, and the exact cubic capacity of the chamber had to be calculated and the appropriate weight of charge prepared for each shot to achieve the same velocity each time. After being withdrawn the barrel was to be sent back to Krupp's works and re-bored and re-rifled at 24cm calibre, then sent back, refitted, and fired with a fresh outfit of shells. Each shell was numbered, and each progressively higher number had a slightly larger diameter of driving band to counter the gradual wear. According to some reports, one shell was loaded in the wrong sequence and blew up the second gun after only a few shots. The first and third mountings continued to fire alternately, renewing their barrels from time to time. In all, 303 shells were fired at Paris, of which 183 landed within the city; 256 Parisians were killed and 620 wounded by the firing. The Allies made great efforts to discover the guns and open fire on them, even sending aircraft to bomb them, but they were well concealed and were never discovered or damaged, being finally withdrawn in the face of the Allied advance in August 1918. Neither the guns nor their mountings were ever found by the Allied Disarmament Commission, and it is not known what happened to them. The concrete emplacements can still be seen in the Forest of Crépy, overgrown but still serviceable.

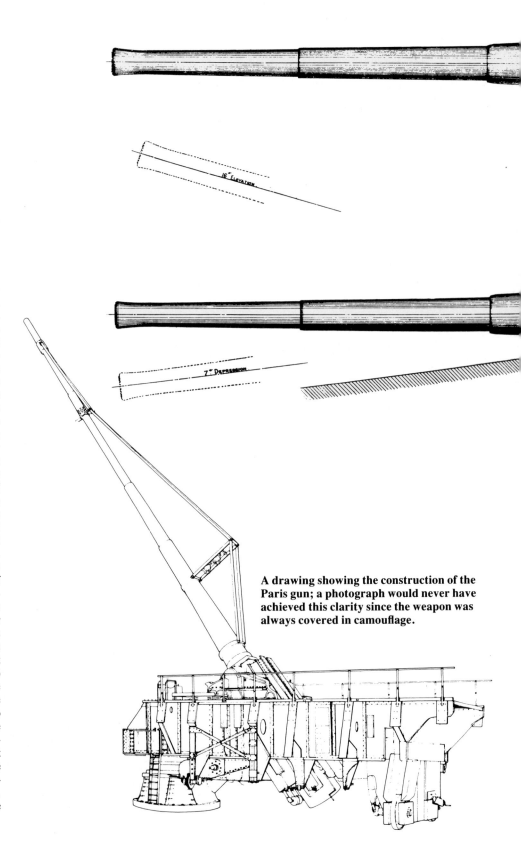

A drawing showing the construction of the Paris gun; a photograph would never have achieved this clarity since the weapon was always covered in camouflage.

PLAN

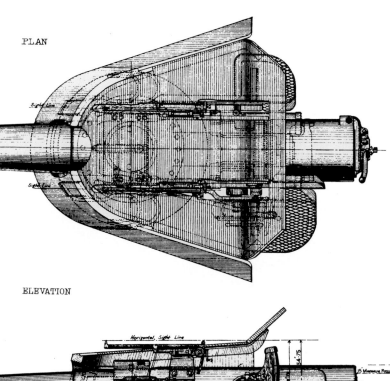

ELEVATION

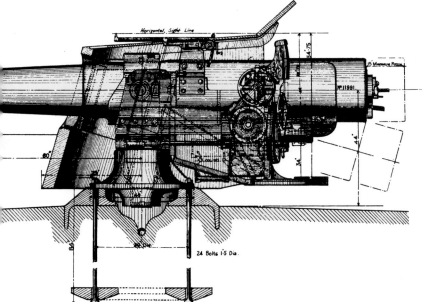

**PEDESTAL MOUNT** A type of mounting for coast artillery guns, it consists of a cast steel pedestal bolted securely to the concrete floor of the emplacement. The top of the pedestal is formed into a socket and may also have a racer (qv) surrounding the socket. A Y-shaped top carriage or yoke fits into the socket, so that it is free to revolve, and if the gun is heavy may be additionally supported by rollers riding on the racer. The arms of the top carriage carry the trunnions, and into these the gun cradle and gun are mounted. The gun thus traverses round the top of the pedestal, and elevation is about the trunnions on the top carriage.

REAR END VIEW.

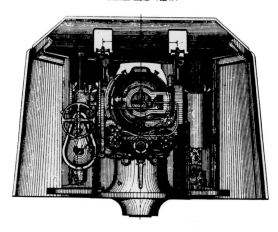

**A British 6in coast defence gun on a pedestal mount.**

**PARROTT** Captain Robert P. Parrott (1804-77) served in the US Army from 1824 to 1836 and then retired to become Superintendent of the West Point Gun Foundry at Cold Spring, New York, in the 1860s. He developed a system of strengthening cast-iron smoothbore guns by shrinking on a hoop of wrought iron around the breech end, then rifling them and so converting them into rifled muzzle-loaders. The Parrott 12pr field gun was used during the Civil War and was quite successful, but heavier guns, developed for naval use, were found to be less reliable. Parrott also developed a number of 'new manufacture' guns, which were based on the same system of shrinking a heavy breech hoop around a wrought-iron barrel, and these were rather more successful than his conversions.

**A group of American Civil War gunners with their 3in Parrott gun. Note the characteristic reinforcing hoop around the breech.**

Above: British troops examining a captured German Pom-Pom in 1916.

The Maxim one-pounder Pom-Pom was little more than an overgrown machine gun.

**PLATFORM** The surface upon which a gun sits during firing. The term is capable of several interpretations; thus the patch of ground upon which a field gun is put into action is often called the 'gun platform'. However, the term is more properly used to refer to a manufactured structure placed under the gun. In the 19th and early 20th centuries, wooden platforms were placed on the ground and siege guns were run on to them for firing, being restrained on the platform by ropes or by hydraulic buffers connected to ground anchors. This kept the gun behind its protective emplacement during siege operations, because without a platform and means of restraint the gun carriage would recoil several feet and expose the gunners to fire while they were restoring it to its firing position.

In more recent years the term invariably refers to a steel platform placed beneath the wheels of a field gun – eg, the 25pr or 105mm Light Gun – or placed centrally beneath the carriage of a heavy gun and connected to the gun carriage to allow the weapon to be traversed over much wider arcs (up to 360° if necessary) than would be possible with the gun wheels on rough ground. In the case of field guns it allows rapid changes of direction to permit them to track and fire at moving targets such as tanks. It makes traversing much easier for heavy guns.

**POINT BLANK** A gun barrel laid horizontal is said to be at 'point blank'. The term comes from the medieval gunner's quadrant (qv), which was marked off in 'points' of elevation. When the barrel was horizontal, the measuring plumb-line aligned with the blank space on the quadrant, hence 'point blank'. Thus 'firing at point blank range' means firing at a range so short that no elevation of the barrel is required.

**POLE TRAIL** The trail of a field gun which consists of a single member mounted centrally on the axle and hence directly beneath the gun. It has the advantage of lightness, which made it popular when guns were horse-drawn, but restricts the amount of elevation as the gun breech will eventually strike the trail as it elevates. This was unimportant until the 19th century because guns were not capable of very great ranges, nor did the tactics of the time demand them, so that the restriction of range due to the small amount of elevation possible was accepted. But as guns became more powerful and fighting ranges increased, the pole trail was gradually abandoned.

**POM-POM** A 37mm calibre automatic cannon devised by Hiram Maxim and put into production in the early 1890s. It was, in fact, an overgrown Maxim machine gun, using the same method of operation and being fed by a canvas belt containing the rounds of ammunition. Originally developed for naval use, as an anti-torpedo-boat weapon, it was placed on a wheeled carriage and was first used as a land weapon by the Boers during the South African War. The rapid arrival of the small 1lb (450kg) shells, and their detonation, gave rise to its nickname 'Pom-Pom', bestowed by British troops. It was later adapted by the British as a light anti-aircraft gun in 1914, but it had insufficient range for this role and fell into disuse. The name, however, was then applied to a number of automatic weapons of similar calibre used by the Royal Navy up to the end of the Second World War.

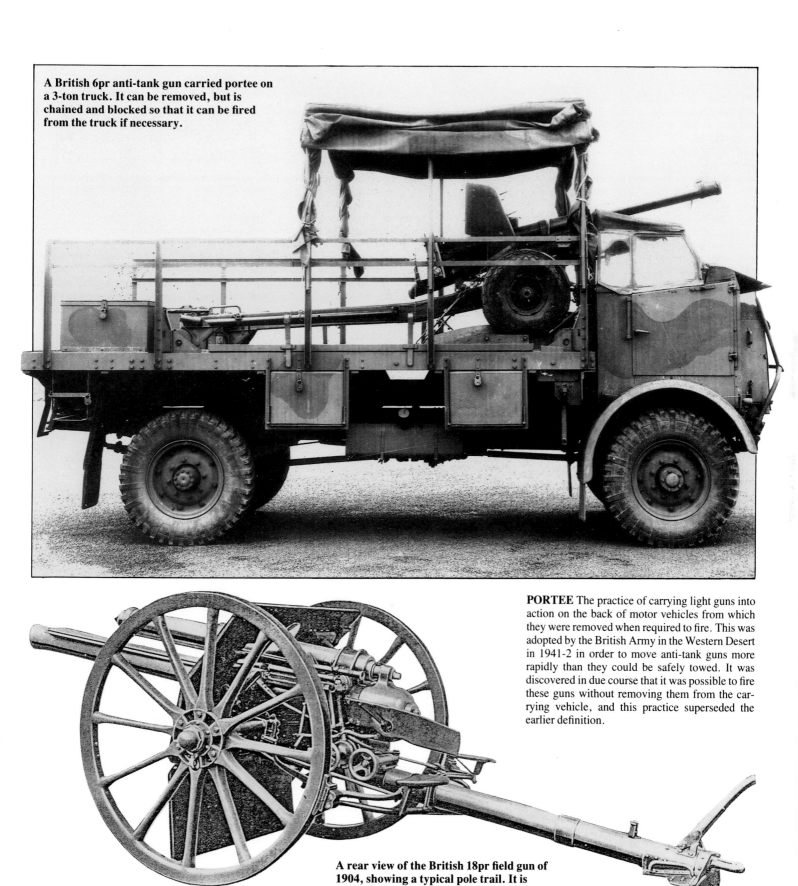

A British 6pr anti-tank gun carried portee on a 3-ton truck. It can be removed, but is chained and blocked so that it can be fired from the truck if necessary.

**PORTEE** The practice of carrying light guns into action on the back of motor vehicles from which they were removed when required to fire. This was adopted by the British Army in the Western Desert in 1941-2 in order to move anti-tank guns more rapidly than they could be safely towed. It was discovered in due course that it was possible to fire these guns without removing them from the carrying vehicle, and this practice superseded the earlier definition.

A rear view of the British 18pr field gun of 1904, showing a typical pole trail. It is obvious from this photograph how the pole trail limited the gun's elevation.

**PREDICTOR** A form of computer which, given the precise position of an aircraft and fed with the subsequent course and speed, would calculate a future position and also calculate firing data for an anti-aircraft gun which would cause the shell to strike the aircraft. The predictor was fed with information from an optical tracking telescope that gave elevation and azimuth of the aircraft, from a rangefinder which gave range, and from an internal clock which gave time. These produced the aircraft's course and speed. It was also given the position of the gun and the ballistic data relating to the shell and fuze in use. Having deduced a future position of the aircraft, it would calculate gun data and compare its result with the future position. If the two failed to agree (ie, if the shell would have missed), it recalculated until it found a gun data solution that would place shell and target in the same place at the same time. This data was electrically transmitted to the gun. Most predictors used during World War Two were mechanical analogue computers; electronic computers did not appear until the closing days of anti-aircraft gunnery. Present-day electronic digital computers are used with a number of self-propelled anti-aircraft gun equipments.

Predictor is the British term; Americans referred to them as 'directors', Germans as 'Funkmessgeräte'.

**A Japanese Type 97 anti-aircraft predictor, a typical 1930s mechanical computer for determining the future position of a target.**

**PREMATURE** The accidental explosion or detonation of a projectile before reaching its intended point of effect. If this takes place along the trajectory, it is known as a 'flight premature'. If it takes place inside the barrel of the gun, it is a 'bore premature'. A flight premature is inevitably an ammunition fault, usually due to a faulty fuze, though in exceptionally long-range guns atmospheric heating of the shell has been known to cause premature detonation of the explosive content. Bore prematures are far more dangerous, because with modern high explosive shells they are likely to blow the gun to pieces and kill or injure the gunners. These are mostly ascribed to ammunition failures, though there are mechanical defects in the gun which can cause prematures; these include steel choke, where the metal in the gun has been displaced and constricts the bore, excessive wear that causes the shell to be checked in its progress up the bore, or the moving of a chamber liner. Ammunition failures causing prematures include faulty assembly or setting of fuzes, air spaces and irregularities in the high explosive filling, degradation of the explosive due to storage in excessively high temperatures, and dirt or grit entering the shell due to carelessness in handling or storage. Modern methods of ammunition manufacture and inspection have made prematures from these causes exceptionally rare, and careful maintenance and inspection of the gun by its commander make gun-related prematures even more uncommon.

**PREDICTED FIRE** Indirect artillery fire in which fire for effect is opened on the target without observation or preliminary ranging. The map location of the target is obtained and gun range and azimuth to this point calculated. This data is then corrected for meteorological conditions, ballistic changes in the gun and other variations from standard, and the gun fired on this corrected data. The rounds should then fall in the target area. First attempted during the First World War, predicted fire has gradually become the normal method of engagement, because it is economic in ammunition and has a greater psychological effect on the enemy, he not having had the usual warning given by ranging rounds. Due to modern use of radar, laser rangefinding, computers and improved methods of determining 'meteor' (qv) corrections, predicted fire is now extremely accurate and, because of the facility for rapid calculation offered by computers, is as quick as observed fire.

**Below: The remains of an American 175mm gun after a high-explosive shell had detonated prematurely in the bore.**

**Above: British troops in an American 105mm Priest self-propelled howitzer; the anti-aircraft machine gun 'pulpit' that gave rise to the name can be seen here.**

**PREPONDERANCE** The imbalance of a gun barrel about its trunnions. In older times when guns were usually fitted with trunnions approximately at their centre of balance, a slight degree of muzzle preponderance was preferred so that the elevating gear was always under stress in the same direction which tended to even out errors. Today, when guns are invariably trunnioned to the rear to allow them to elevate without the breech striking the ground, muzzle preponderance is considerable but is counteracted by balancing gear. See 'Equilibrator', 'Counter-weight'.

**PRIEST** Name given by the British to the American 105mm self-propelled howitzer M7, reputedly because the anti-aircraft machine gun cupola resembled a pulpit. The M7 was the standard M2A1 field howitzer mounted in an open-topped compartment and built on the chassis of the M3 'Grant' tank. Development began in 1941 and it was standardized in November 1943, though it entered service before that time, the first being used by the British at the battle of El Alamein in October 1942. Priests remained in British service until replaced by the 25pr 'Sexton' (qv), while those in US service remained until replaced by the M37 in 1945. A total of 3,500 were built and many were employed by other armies until the 1950s.

It is worth recording that the first British self-propelled gun was known as 'Bishop', a randomly-assigned code-name. When 'Priest' was taken into common use, the Royal Artillery, without formal agreement or record, adopted the habit of using ecclesiastical names for their self-propelled equipments, leading to 'Deacon', 'Sexton' and 'Abbott'. 'Archer' and 'Achilles' were exceptions to this rule because they were tank destroyers and not field artillery equipments.

**PROBABLE ERROR** That distance around the predicted point of impact in which 50 percent of the shots fired will fall.

**PROBERT RIFLING** A form of gun rifling developed for high velocity guns by Colonel Probert of the British Armament Research Establishment during the Second World War. The essential feature of Probert rifling was that the depth of the rifling groove gradually reduced as the rifling neared the gun muzzle, until at a point about six calibres from the muzzle the bottom of the groove ran out level with the top of the lands and thus the gun was smoothbored from there to the muzzle. This method was used in conjunction with a special driving band which was pressed into a groove on the shell as the rifling depth decreased, and was finally smoothed flush with the shell body during its passage of the smoothbored portion. The object was to develop a high muzzle velocity by ensuring a very tight seal between driving band and rifling and also to remove any excrescences on the shell which might cause drag during flight. It was used with the British 3.7in Mark 6 anti-aircraft gun.

**Above: One of the few surviving Armstrong protected barbette emplacements, at Packpool Fort, Isle of Wight. The gun mounting revolved on the central drum and the barrel depressed so that the gun could be muzzle-loaded under cover of the pit wall.**

**Inset: Proving a gun is not a mere bureaucratic formality; things sometimes go wrong as here, where a flaw in the steel caused the barrel to split.**

**Left: An early form of gun proof, employed in the 16th century, was to attack the gun with hammers in order to detect a faulty casting prior to firing.**

**PROTECTED BARBETTE** A system of emplacing heavy muzzle-loading coast defence guns, developed in the 1870s by Sir William Armstrong. The emplacement was a semi-circular parapet of concrete, and the gun carriage was mounted to pivot on the centre of the semi-circle. After firing, it could be traversed to one side or the other, as was convenient, and depressed so that the muzzle was aligned with an aperture in the emplacement. Inside this aperture was the loading apparatus, consisting of lifts to bring up shells and cartridges and hold them opposite the gun's muzzle, and a rammer, usually operated by steam. With the gun muzzle aligned, the cartridge was rammed, followed by the shell. The gun could then be elevated and traversed back to the firing position. The purpose was to allow the gun to be loaded without exposing men to return fire, as would happen if the gun was loaded in the open, which was customary at that time. Two Armstrong Protected Barbettes are known to remain, one in Malta and one in Gibraltar, together with the 100-ton 17.75in guns mounted in them, and an empty emplacement can be seen at Puckpool Battery on the Isle of Wight. It is possible that others may still exist in former British colonies.

**PROOF** The testing of a gun after manufacture and before issue for service. The gun is first carefully examined to ensure that it complies with the specifications, after which it is fired with a propelling charge and a special heavy shot which, in combination, give a chamber pressure in excess of that expected in service by a specified percentage – usually 10-15 percent. The firing is conducted 'under precautions', which means that the gun is loaded, all personnel removed to a safe place behind cover and the gun fired by remote control. Several proof rounds of this type are fired, after which the gun is again carefully examined to ensure that the barrel has not expanded, the rifling is not damaged and that, all in all, it does not show any signs of distress. If all is correct, the gun is stamped with a proof mark and certified as being fit for service.

# Q

**QUADRANT ELEVATION** The angle between the horizontal plane and the axis of the bore when the gun is laid, before firing. It consists of tangent elevation (qv) and angle of sight (qv). The name is derived from the Gunner's Quadrant (qv).

**QUARTER-TO-TEN GUN** British soldier's nickname for the 9.45in Skoda Siege Howitzer M1898, eight of which were purchased by Britain in 1900 for use at the Siege of Pretoria. They arrived too late, however, and were never used in South Africa. Two were sent to China in 1902, though for what purpose is unknown. All were returned to England in due course and relegated to a training role. They were declared obsolete in 1920, never having been used in action. Weight in action: 8.5 tons (8,635kg); barrel length 92.5in (235cm); shell weight 280lb (128kg); muzzle velocity 928ft/sec (285m/sec); maximum range 7,650 yards (7,060 metres).

**QUICK LOADING GEAR** Apparatus built into the elevating mechanism of a gun (or more usually a howitzer) which allows the barrel to be disconnected from the elevating mechanism and brought to the horizontal very quickly for re-loading, then re-elevated and reconnected to the elevating mechanism. This permits the laying of the gun to continue during loading and saves the gunlayer the task of winding the gun barrel down and up again before returning to his laying. Seen, for example, on the British 5.5in medium gun and on the US 12in Seacoast Mortar M1912. See also 'sleigh'.

# R

**RACER** Steel arc or circle let into the surface of an emplacement or pedestal and upon which the gun mounting rotates on wheels or rollers.

**RACER CORRECTIONS** Corrections applied to a gun of position – eg, a coast defence or static anti-aircraft gun – to compensate for any lack of level of the racer (qv) due to subsidence of the emplacement floor. A racer out of level would cause the gun barrel to vary in elevation as it traversed; by measuring the irregularity it is possible to deduce corrections to elevation that need to be applied at certain angles of azimuth.

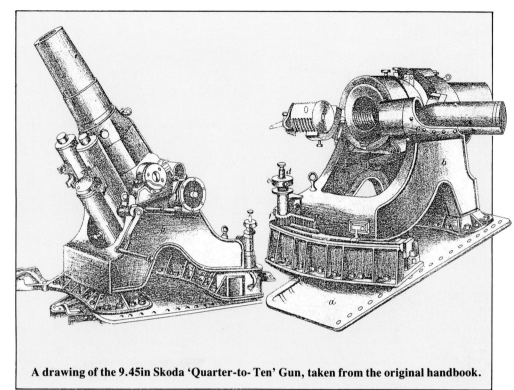

A drawing of the 9.45in Skoda 'Quarter-to-Ten' Gun, taken from the original handbook.

**Inset: An old fortress gun emplacement, showing the front and rear racers and the pivot on which the gun carriage traversed.**

**This muzzle-loader used by Union troops in the American Civil War was one of the first railway guns.**

**RAILWAY GUN** Any piece of artillery mounted on a carriage intended to run on conventional railway lines.

Mounting on railway wheels permits large guns to be moved relatively easily and quickly over long distances, and as a result the railway gun has always been popular with Continental powers which might wish to move their artillery reserve from one side of the country to another very quickly. The only restriction is the loading gauge of the railway, and the gun must be designed to pass this, though in some cases the gun has one configuration for travelling and another for firing; in extreme cases the mounting is carried in individual components and assembled on site, though this is not strictly a railway gun.

Calibres of railway guns range from 150mm (6in) at the lower end to 520mm (20.5in) and even more at the upper end. In the smaller calibres the gun is usually mounted to traverse through 360°. The mounting is stabilized by struts or outriggers extended at the side of the track. As the calibre increases so does the size of the mounting, and it soon becomes impracticable to permit all-round traverse. In the largest calibres the gun has no traverse whatever inside the mounting, and changes in the direction of fire can only be done by moving the mounting along a curved track or 'épée' (qv), mounting it on a turntable, or using winches to pull the front end of the mounting sideways across its railway truck units. An alternative method, rarely seen, is to lay a T-shaped piece of track, and by means of jacks, arrange the mounting with its foretruck on the cross-bar of the T and it rear truck on the upright; the foretruck can then be moved across the cross-bar to give changes of direction, the mounting pivoting on the rear truck.

Recoil can be suppressed by conventional recoil systems fitted between the gun and the mounting; by dual recoil systems (qv); or by trunnioning the gun directly into the mounting and relying upon its enormous weight to damp down the recoil. In the heavier guns the mounting itself is left free to move back along the track to absorb some of the recoil force, and is checked by conventional brakes; it is returned to its firing position by winches or by a locomotive. Other types of mounting were arranged so that the body of the mounting was lowered on to either the track or a prepared platform assembled alongside the track, and the wheels relieved of weight. Ground anchors prevented the mounting from moving. Another system was to lower the mounting to the rails or to a prepared platform, but allow it to slide backwards under recoil; in this case it became necessary to jack it back on to its wheels after every few shots and run it back to its starting position. It should be said that the old soldiers' tales of guns recoiling into tunnels for concealment after each shot, or guns that recoiled down the line to the nearest station and brought up the mail and rations on run-out, are entirely without foundation.

**Italian gunners loading a 152mm gun. The shell is being inserted into the breech. One man stands by with the rammer.**

**RAMMER** A device used to force the shell into the chamber and ensure that it grips either the rifling or the forcing cone and is in the position laid down by the gun designer to enable the correct pressure to be developed in the chamber. The rammer may be nothing more complicated than a long pole with a suitable head, pushed behind the shell by one or more men, or it may be a complex spring or hydraulic-driven piston assembly that does the same thing but in a more regular and precise manner.

**RANGING** Ascertaining the range and azimuth required to hit the target by firing a series of trial shots and making corrections to bring them on to the target. Known also as 'adjustment'.

**RATEFIXER** A British development project intended to develop high-speed loading systems to permit high rates of fire with medium-calibre anti-aircraft guns. The project began in 1946 and ran until 1949.

Mechanical loading systems had hitherto always sought simply to duplicate the movements of hand-loading: to present the round of ammunition to the breech, ram it, close the breech, fire, open the breech, eject the spent case, and present the next round. Ratefixer suggested that this might not necessarily be the right answer; it might, for example, be possible to begin presenting the next round while the gun was being fired, and arrange the paths of the incoming round and the ejected case so that they passed each other in mid-flight. Four different designs were built: Ratefixer K, designed by Captain Kulikowski of the Polish Army used two drums at 90° to the mounting, feeding alternating arms which swung the cartridges across to a central rammer; Ratefixer C, by Colonel Carmichael of the Royal Electrial and Mechanical Engineers, used a 30-round hopper feeding a cross-table and through

a five-foot diameter trunnion; Ratefixer CR by Russell Robinson, a machine gun designer, was the 'C' model modified to be belt fed; and Ratefixer CN, by the Frazer-Nash company, was belt-fed and powered by an hydraulic motor.

The experimental weapons all worked well and 'CN' eventually reached a rate of fire of 75 rounds per minute, which meant moving 2.75 tons of cartridges per minute into the gun. The lessons learned in Project Ratefixer were then applied to the development of new anti-aircraft guns, notably 'Longhand' (qv) and 'Green Mace' (qv), but in the event the guided weapon came to fruition and neither of these weapons entered service.

**RAUSENBERGER** German ordnance engineer and chief designer for Krupp in the period 1890-1918, Professor Rausenberger was responsible for innumerable designs, but perhaps his most famous were 'Big Bertha' and the Paris Gun.

**RECOIL** The rearward motion of the gun due to reaction to the ejection of the projectile from the muzzle. In broad terms, the amount of recoil is proportionate to the ratio of shot weight to gun weight, so that the larger the gun the less, generally, is the recoil.

**RECOILLESS GUN** A gun in which recoil (see above) is absent, due to rearward motion of a 'counter-shot' of similar momentum to that of the shell. In its simplest form, the Davis recoilless gun, there are two barrels back-to-back on a common chamber. On firing, the front barrel ejects a shell of X weight, while the rear-facing barrel ejects a mass of grease and lead shot also of X weight and at the same velocity. As a result, the action and reaction are equal and the gun does not recoil.

It can be seen that by halving the weight of the counter-shot and doubling its velocity, the momentum remains the same and therefore the gun would

**RECIPROCATING SIGHT** Any sight that is capable of being cross-levelled (qv) to compensate for any lack of level of the trunnions.

**The German 75mm LG40 recoilless gun, used by parachute troops in the assault on Crete in 1941.**

An advantage of the recoilless gun is that its light weight allows a powerful weapon to be carried on a light vehicle – in this instance a 106mm RCL gun on a British Land Rover.

still be in balance. Taking this to a practical conclusion, we arrive at the modern recoilless gun where the ejected shell is counter-balanced by a mass of extremely high velocity gas of low weight but still having the same momentum. The gas is produced from the propelling charge, which is of much larger weight than would be required in a conventional gun since about four-fifths of it go to produce the gas required for the counter-balancing. The gas is allowed to leave the gun chamber under control and is directed rearwards through a venturi to increase its exit velocity.

The recoilless gun can be extremely light for a given calibre because it does not require a recoil mechanism (invariably heavy) nor does the carriage need to be very substantial. Its drawbacks are that it cannot develop a very high velocity, and there is a considerable danger space behind the gun into which the counter-balancing gas is discharged at high velocity and temperature. This discharge also gives the weapon a very prominent 'firing signature', which means that it is easy to spot on the battlefield as soon as it opens fire.

**RECOIL MECHANISM** A mechanism designed to absorb some or all of the gun's recoil and return the gun into battery after recoil has been completed. It is usually interposed between the gun and the carriage by being attached to the cradle so that the gun can move and operate the system, which then

acts as an elastic link between the gun and its mounting.

Guns had no recoil system until the 1880s. If wheeled, they bounded back across the ground when fired and had to be pushed back into their firing position after every shot; it is said that at the end of the battle of Waterloo the British gunners were so exhausted that they were unable to man-handle the guns back into position and were gra-dually moving backwards across the battlefield. The first attempts to absorb recoil were made in guns of position in fortresses, where the gun was attached to a carriage which was permitted to recoil across a 'slide', an inclined plane carried on wheels and pivoted to give direction to the whole mount-ing. The gun and carriage recoiled up the slide on firing and came to a halt due to the friction between the two units, often artificially increased by throw-ing sand on the surface of the slide. After reloading, the gun and carriage were pushed forward, down the slide, by means of handspikes and levers, or by levering two small wheels into place to allow the friction to be released and permit gravity to assist the forward movement.

This system was later improved by using the 'compressor' (qv), a system of interleaved plates that braked the rearward movement but which could be thrown free to permit the return of the gun and carriage. The hydraulic brake was introduced in the late 1870s at the suggestion of Professor Siemens, a British engineer. In this system, an

oil-filled cylinder was attached to the slide, and a piston-rod to the carriage. The piston rod was pulled through the cylinder as the gun and carriage recoiled. Holes in the piston head allowed oil to flow from one side of the piston head to the other, and acted as a brake to the gun's movement. The size and number of holes were determined for each different type of gun by a process of trial and error.

The idea of using the hydraulic brake on more compact types of mounting evolved from Professor Siemens' suggestion. Rotary or other valves were used to alter the amount of resistance so that the recoil movement was gradually damped down, springs were added that compressed during the recoil movement and expanded afterwards to thrust the gun back into battery.

The first wheeled gun to adopt an hydraulic system with any degree of success was the French M1897 75mm gun, which used a compact system containing an oil buffer to regulate the recoil stroke and a nitrogen-filled 'recuperator' cylinder to thrust the gun back into battery. The recuperator cylinder held the gas under pressure. This pressure was increased by the movement of the gun during recoil acting on a separate piston to force the gas into a smaller space. The French were able to conceal the exact system of operation of this mechanism until the First World War, but by the early 1900s several other engineers had developed comparable systems and the use of recoil mechanisms had become commonplace by 1914.

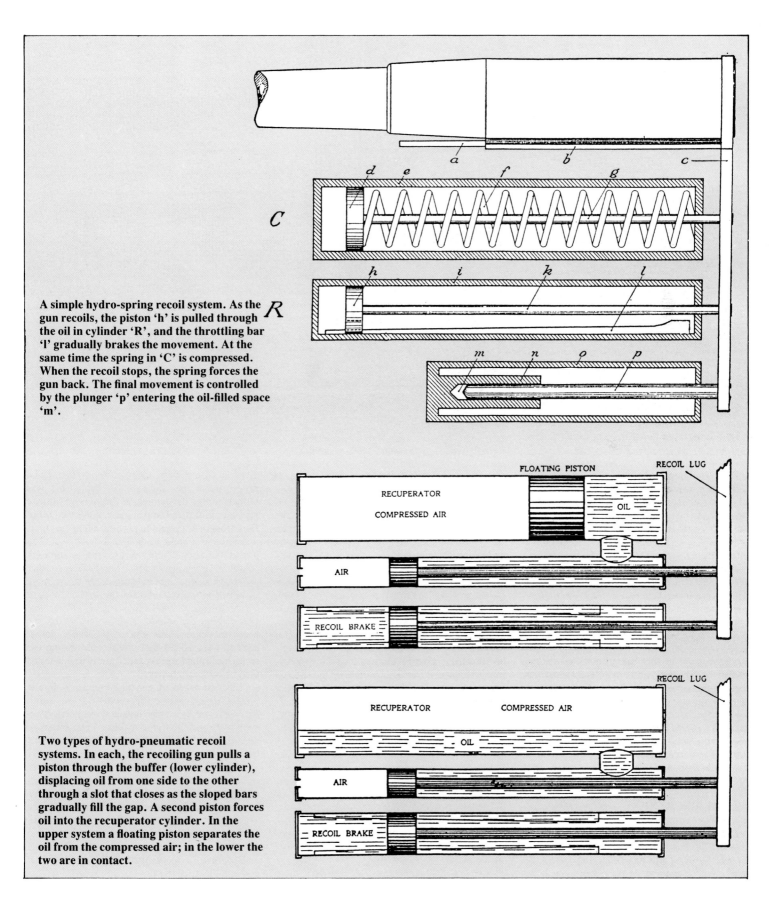

A simple hydro-spring recoil system. As the gun recoils, the piston 'h' is pulled through the oil in cylinder 'R', and the throttling bar 'l' gradually brakes the movement. At the same time the spring in 'C' is compressed. When the recoil stops, the spring forces the gun back. The final movement is controlled by the plunger 'p' entering the oil-filled space 'm'.

Two types of hydro-pneumatic recoil systems. In each, the recoiling gun pulls a piston through the buffer (lower cylinder), displacing oil from one side to the other through a slot that closes as the sloped bars gradually fill the gap. A second piston forces oil into the recuperator cylinder. In the upper system a floating piston separates the oil from the compressed air; in the lower the two are in contact.

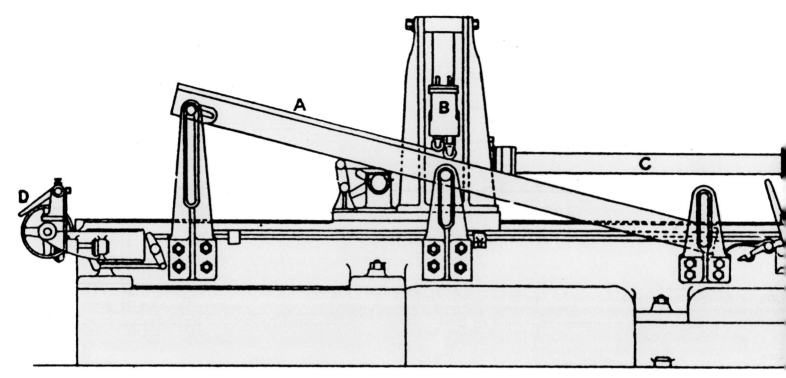

**RECUPERATOR** The part of the recoil mechanism (see above) that returns the gun to battery after recoil is completed. It can be a spring or bank of springs compressed during the recoil stroke or, more common today, a charge of compressed gas held in a cylinder in the recoil mechanism and further compressed during the recoil movement. When recoil stops, the gas pressure forces on a piston which, by various means, transfers the effort to the gun and thus pushes the gun back into battery.

**REGISTRATION** Ascertaining or checking the firing data for targets by actually firing a single gun at them. This is usually done to prepare data for a fire plan for use in the near future. Thus, target data some days old may be taken, present 'meteor' (qv) and other corrections applied and, before ordering this to all the guns, registration by a single gun is carried out to check that the calculations are correct.

**REPLENISHER** An oil tank attached to the gun carriage to keep the recoil system topped up with oil during prolonged firing.

**RETRACTING GEAR** Winching device used with disappearing guns and with some barbette carriages to pull the gun back to the fully recoiled position without firing. It may be done to exercise the recoil system and ensure it is functioning correctly, or to place the gun in the loading position prior to beginning an engagement. The term is also applied to systems for pulling back the barrels of large field equipments to place the centre of gravity over the roadwheels for easier towing and to support the barrel against vibration during travelling. In cases where the purpose is to exercise the

recoil system without firing, the gear is sometimes called a 'gymnasticator'.

**RHEINMETALL** German arms manufacturer that has its principal factory at Düsseldorf. Founded as the Rheinische Metallwaaren- und Maschinenfabrik in 1889 to manufacture ammuntion for the Mauser rifle, the key figure in the company's success was Dr. Heinrich Ehrhardt. He began designing light artillery, and the British adoption of his 15pr QF gun in 1901 was a major milestone in the company's progress. After the First World War the company remained in the heavy engineering field but formed alliances with companies outside Germany which enabled it to continue designing and developing weapons for manufacture abroad to avoid the restrictions of the Versailles Treaty. Rheinmetall began developing artillery for the German Army in the 1920s and its 105mm light field howitzer became the standard German divisional weapon. The company merged with the Borsigwerke in 1936, a well-known firm in the heavy industrial and locomotive field, to become Rheinmetall-Borsig AG. By the end of the Second World War the company was operating 12 major factories and was manufacturing every type of munition for all types of weapon from machine guns to the heaviest types of artillery. The factories, many of them severely damaged, were closed down after 1945 but the company remained in business and in the 1950s returned to military production by building the wartime MG42 machine gun for the newly-formed Bundeswehr. They then became the German partner in the tripartite consortium building the FH70 155mm howitzer.

**RICOCHET FIRE** Fire by field artillery guns at a low angle and using a delay fuze so that the shell strikes the ground some distance in front of the target and bounces off into the air. The strike is sufficient to initiate the action of the fuze, and the delay then functions as the shell is 'on the bounce' to burst it in the air over the target. It is used as a method of obtaining airburst without resorting to the use of time fuzes, and is common practice in the French and US armies, but rarely in the British.

**RIFLED BREECH-LOADER** Although almost all modern guns are rifled and breech-loading, this particular phrase and its associated abbreviation 'RBL' are specifically used by historians to refer to the Armstrong breech-loaders developed in the 1850s.

**RIFLED MUZZLE-LOADER** A rifled gun that is loaded at the muzzle, this type of weapon was used between 1850 and 1890 by several countries due to the difficulty of developing a breech closing system strong enough to withstand the heavy charges needed for coast defence, naval and siege guns and not involving some patented design feature. The British Armstrong RML guns were typical, being built-up guns using three deep spiral grooves as the rifling. The shells had three rows of soft metal studs that were engaged with the rifling when loading and imparted the required spin by riding on the rifling when fired.

The RML principle is not entirely dead; there are a number of medium-calibre (100-120mm) infantry mortars that still use it, though instead of studded projectiles other methods of rotating the shell are used.

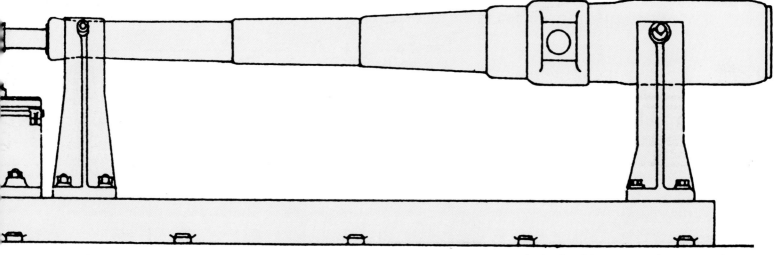

**RIFLING** The helical grooves cut into the interior surface of the gun in order to impart spin to the projectile.

Spinning the shell about its longer axis gives it gyroscopic stability and keeps it pointing towards the target. In modern guns the spin is imparted by a soft metal 'rotating' or 'driving' band which is pressed into the shell body; this band bites into the rifling grooves, under the pressure of the cartridge explosion, and is attached to the shell in such a way that the rotary motion is imparted to it.

The number of grooves depends upon the calibre; a rule of thumb is six grooves for every inch (25mm) of calibre, though this is not rigidly adhered to. The width, depth and spacing of the grooves really depends upon the type of shell being fired, the maximum pressure and the degree of twist of the rifling, and is a matter for the gun designer to decide.

The twist of the rifling can either be 'uniform', in which case it is cut on a regular curve from the chamber to the muzzle, or it can be 'increasing' or 'parabolic' or 'gain twist', in which case it starts with a very small amount of inclination and gradually becomes steeper as it approaches the muzzle. The object of increasing twist is to allow the shell to gain forward motion before applying rotary motion; guns with high chamber pressures place a high stress on the driving band at the instant it begins to engrave, and making it rotate suddenly could possibly cause the driving band to fail by shearing. Some guns have had their grooves straight for almost a third of the bore before developing the required amount of twist.

The amount of twist is decreed by the size of the shell; generally speaking the longer the shell, the

A rifling machine; the cutter is pushed into the gun by the boring bar 'c' driven forward by the moving carriage 'b'. An arm on the boring bar follows the profile bar 'a', which turns the cutter and so traces out a single grove inside the gun.

more twist it needs, but there is a point at which it becomes impossible to develop a stablizing spin; this is when the length of the shell is more than about seven or eight times its calibre. There are, though, other factors, particularly with modern guns firing specialized ammunition.

The amount of twist is specified in various ways. The system used in British and American service is to describe the rifling as having one turn in so many calibres (eg, one turn in 30 calibres), which means that the shell will make one complete revolution in a distance thirty times the calibre of the gun. This, obviously, can only apply to uniform rifling. In the case of increasing twist it is generally sufficient to say that the rifling commences with a turn of one turn in X calibres, increases to one turn in Y calibres at a point Z inches from the muzzle, and thereafter remains uniform. Continental practice is to describe the pitch of the rifling grooves by the angle the groove makes with the axis of the bore; eg, 5°45'. Again, an increasing twist would be described as starting at X° and increasing to Y°. It is, of course, the pitch of the rifling at the muzzle that decides the rate of spin of the shell.

**ROBINS** Credited with 'founding the science of gunnery', Benjamin Robins (1707-51) was a mathematician and experimenter who devoted much of his life to investigating the performance of artillery. He invented the ballistic pendulum (qv) and developed various mathematical equations for

determining range, elevation and velocity that were to be used for another century before better methods were devised. His most important work was *New Principles of Gunnery*, published in 1742, which displayed some very advanced theories for its time and was the inspiration for most experimental artillery work for the next century. His most famous quotation is perhaps one on the subject of rifling, which in Robins' day was scarcely thought of: 'Whatever state shall thoroughly comprehend the nature and advantages of rifled barrel pieces and . . . shall introduce into their armies their general use . . . will by this means acquire a superiority which will almost equal anything that has been done at any time by the particular excellence of any one kind of arm . . .' Appointed Engineer-General to the East India Company, he went to India to make recommendations on the fortifications there, was taken ill, and died at Fort St David, Madras.

**ROCKING BAR SIGHT** Commonly used with artillery, it consists of some form of sight – optical or otherwise – mounted in such a way that it can be moved by the movement of the gun in elevation but also moved independently of the gun. To operate the sight the required elevation or range is set on a scale on the rocking bar and the bar is thus moved so that the sight line is depressed. The gun is brought to the elevation required by elevating the gun until the sight line is laid on the target. To cater for indirect fire, the 'line of sight' is replaced by a level bubble.

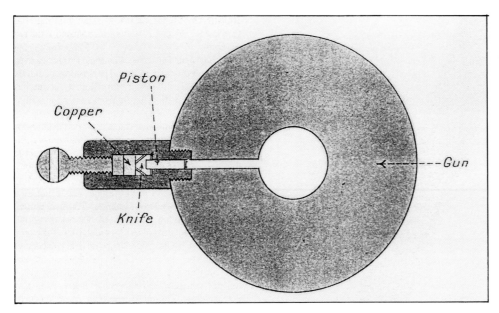

The Rodman pressure gauge; gas pressure inside the gun forces out the piston driving the knife into a slug of copper. The resulting indentation is proportional to the internal pressure.

**RODMAN** Captain Thomas J. Rodman (1815-71), US Army, developed a system of casting ordnance in the 1860s in which the core of the mould was water-cooled. The inner surface of the barrel cooled rapidly and solidified, while the remaining metal, which cooled more slowly, placed the inner portion under compression due to contraction during the cooling period. This placed a compressive stress on the surface of the bore and improved the gun's resistance to the explosion of the propelling charge. Rodman guns were manufactured in several calibres but were principally produced in large calibres for naval and coast defence use, the largest being of 20in bore. Numbers of 15in guns remained in place in coast defences until the 1890s. The Rodman gun is generally considered to be the high-water mark of smoothbore muzzle-loading artillery design and manufacture.

Rodman's name also attaches to a pressure gauge that he invented for determining the pressure of the explosive gas inside the gun. It consisted of a steel cylinder containing a piston and a carefully-dimensioned pellet of copper, which was screwed into a hole drilled in the gun at the required place. When the gun was fired the pressure of gas entered the steel cylinder and forced a knife blade on the piston to cut into the copper slug. The copper slug could then be examined by removing the gauge and dismantling it, and since the malleability of copper was known, the depth of the cut could be used to determine the amount of gas pressure to which it had been subjected.

**ROLLING RECOIL** A form of recoil control adopted in railway artillery mountings in which the entire gun mounting is permitted to roll backwards on its wheels under the recoil force; it may be braked or allowed to come to a stop by its own inertia. The

gun may then be re-laid and fired from the position in which it has been stopped, or it may be moved back to its original position by a locomotive. In spite of legends, there is no known case where the gun has been able to recoil with sufficient force to 'recoil into a tunnel'.

**ROTATION OF THE EARTH** A factor to be considered in long-range firing, particularly when the projectile passes through the stratosphere. During the shell's flight the earth is spinning at its normal rate, but the gyroscopic stabilization of the shell tends to keep it on a straight line in space, so that, in effect, the earth has moved beneath it by the time the shell comes down on its target. It is therefore necessary to make a correction to the calculated data upon which the gun is fired. Looking at it another way, for a long-range gun every target is a moving target and the gun has to be aimed off accordingly, just as you would when firing a shotgun at a flying pigeon.

**ROYAL ORDNANCE** The British government organization for the manufacture of munitions. The system had its beginning in Woolwich Arsenal (qv), but Royal Ordnance Factories were first built during the First World War in order to cope with the enormous demand for munitions, which far exceeded what could be produced by private contractors. Most were closed down after the war but a handful were kept in an operating condition, and in 1937-41 more were constructed. By the end of the Second World War there were 44 ROFs, operated by the Ministry of Supply, of which about half were concerned with the making of artillery weapons and ammunition. Again, many were closed down and dismantled after the war. There are currently 14 factories which now form Royal Ordnance plc.

**RUNNING BACK STOP** A metal plate or block that can be bolted to the cradle of a gun to prevent any rearward movement of the gun in the cradle. It is a tool used when repairing or adjusting the recoil mechanism to ensure that the gun cannot accidentally move if the recoil system pressure is relieved or reduced. The design usually prevents the breech being opened to ensure that the gun cannot be fired should the stop be inadvertently left in place.

**RUNOUT** The movement of the gun to return it to the firing position ('into battery') after recoil. In modern equipment this is performed by the recuperator (qv); in 19th century carriages it was often done by gravity, the gun having recoiled up an inclined plane.

**A German 105mm mountain howitzer with a running-back stop bolted to the cradle to prevent movement of the gun while the recoil system was being repaired. The 'stop' consists of a simple welded steel bracket, painted red for easy identification.**

**The first Soviet designed artillery piece was this 76mm Infantry Gun Model 1927.**

**The 122mm howitzer M1938 is still in wide use throughout the world; this is the Chinese-manufactured version, known as Type 54.**

**The 57mm Model 1943 anti-tank gun, with a longer barrel and more power than Western guns of the same calibre.**

**RUSSIAN ARTILLERY** Russian gun manufacture in the 19th century was concentrated at the two arsenals of Putilov, near St Petersburg, and Obuchov near Moscow. The guns produced there were of mediocre design and performance, and the production facilities were insufficient, so much of the Russian artillery came from foreign manufacturers such as Krupp, Armstrong and Schneider.

Gun design and manufacture practically came to a stop for several years after the 1917 revolution, and the Red Army relied entirely upon those guns that had been in existence in 1917, plus a number acquired by capture or purchase from various sources. It was not until the late 1920s that guns were again produced in volume. The first models were modernized versions of Tsarist weapons. These can be recognized by their titles; for example, the 76mm Gun M02/30 implies the 1902 design modernized in 1930. This modernization was generally aimed at improving the range, which was usually done by designing new carriages to give greater elevation and developing more powerful ammunition.

**The 122mm D-74 field gun first appeared in 1955; note the folded-up firing pedestal beneath the barrel.**

The 37mm anti-aircraft gun Model 1939 was copied from a Bofors original.

Above: The 76mm M1939 field gun, seen here in its German form as the 76mm PaK 39(r) anti-tank gun. So many were captured in 1941 that the Germans found it worth while to manufacture ammunition and adopt them for their own service. This one was captured by US troops in North Africa.

The 85mm Model 1939 anti-aircraft gun; the shields were added when it was adopted for anti-tank fire, a result of experience with the German 88mm gun in the same role.

Once these designs were completed, work began on a completely new range of artillery, and 76mm guns and 122mm howitzers appeared at various times during the 1930s. Severe losses of artillery immediately after the German invasion of 1941 led to more new designs, mostly developed with an eye to rapid and simple production and robustness in service, and these began to come into use early in 1942. By the latter part of the war the 76mm field gun was gradually being replaced by 85mm weapons, which were more powerful and had a more effective shell.

Heavier artillery in the 122-152mm range was widely adopted, and there were also large numbers of 203mm howitzers, notable for being mounted on a tracked carriage for ease of movement through mud and snow.

Heavy railway and coast defence artillery were also developed, but outside Russia very little is known about these weapons. Anti-tank and anti-aircraft gun development paralleled that of other countries, the calibre and power of the guns increasing in step with improvements in the performance of tanks and aircraft.

Almost the entire artillery inventory has been overhauled since the Second World War, some calibres more than once, although there are still large numbers of wartime guns in stock. New designs appear from time to time, and it is generally some years before full details are known in the West. Moreover, it is not unknown for strange guns to be paraded across Red Square on anniversary days and never be seen again. These are probably intended for nothing more than to give Western Intelligence officers something to think about! The following table lists the principal Russian and Soviet artillery pieces used during this century and of which details are known.

## RUSSIAN/SOVIET ARTILLERY EQUIPMENT

| Equipment | Date | Calibre (mm) | Barrel length (cals) | Breech mech. | Elevation max (deg) | Traverse (deg) | Weight in action (kg) | Shell weight (kg) | Muzzle velocity (m/sec) | Range, max (metres) |
|---|---|---|---|---|---|---|---|---|---|---|
| **Anti-tank artillery** | | | | | | | | | | |
| 45mm M1932 | 1932 | 45 | 46 | VSB | 25 | 60 | 510 | 2.1 | 760 | 8900 |
| 45mm M1937 | 1937 | 45 | 36 | VSB | 25 | 60 | 508 | 2.5 | 765 | 8900 |
| 45mm M42 | 1942 | 45 | 66 | VSB/SA | 25 | 60 | 570 | 2.1 | 820 | 8900 |
| 57mm ZIS-2 | 1943 | 57 | 73 | VSB/SA | 25 | 56 | 1150 | 3.8 | 990 | 8400 |
| 57mm CH-26 | 1955 | 57 | 73 | VSB/SA | 35 | 54 | 2100 | 3.1 | 1000 | 8400 |
| 100mm BS-3 | 1944 | 100 | 56 | VSB/SA | 45 | 55 | 3455 | 15.8 | 900 | 20000 |
| 100mm T-12 | 1955 | 100 | 54 | VSB/SA | 40 | 55 | 2700 | 15.7 | 900 | 21000 |
| **Anti-aircraft artillery** | | | | | | | | | | |
| 37mm Gun M39 | 1939 | 37 | 90 | VSB | 85 | 360 | 2000 | 0.7 | 960 | 6000 ceiling |
| 57mm M50 | 1950 | 57 | 73 | | 90 | 360 | 4000 | 2.7 | 975 | 4000 ceiling |
| 76.2mm M15 | 1915 | 76.2 | 30 | VSB | 75 | 360 | 10160 | 6.5 | 588 | 5500 ceiling |
| 76.2mm M31 | 1931 | 76.2 | 55 | VSB | 82 | 360 | 2750 | 6.5 | 815 | 9500 ceiling |
| 76.2mm M38 | 1938 | 76.2 | 55 | VSB/SA | 82 | 360 | 4300 | 6.5 | 815 | 9500 ceiling |
| 85mm M39 | 1939 | 85 | 55 | VSB/SA | 82 | 360 | 4300 | 9.2 | 800 | 8280 ceiling |
| 85mm M44 | 1944 | 85 | | VSB/SA | | | 4890 | 9.2 | 900 | 10200 ceiling |
| 10cm M49 | 1949 | 100 | 60 | VSB/SA | 80 | 360 | 10665 | 15.8 | 945 | 10650 ceiling |
| **Mountain artillery** | | | | | | | | | | |
| 76mm Gun M04 | 1904 | 76.2 | 13 | SQF | 35 | 5 | | 6.5 | 295 | 4160 |
| 76mm Gun M09 | 1909 | 76.2 | 17 | SQF | 28 | 5 | 625 | 6.5 | 730 | 6080 |
| 76mm Gun M38 | 1938 | 76.2 | 23 | | 70 | 10 | 785 | 6 | 495 | 10100 |
| 76mm Gun M69 | 1969 | 76.2 | | HSB | 65 | 50 | 780 | 6.5 | | 11000 |
| 10cm Howitzer M09 | 1909 | 105 | 10 | SQF | 60 | 5 | 730 | 12 | 300 | 6000 |

The 122mm D-30 howitzer has obvious affinities with the Skoda-designed German leFH 43 design.

*Russian/Soviet Artillery Equipment (Cont)*

| Equipment | Date | Calibre (mm) | Barrel Length (cals) | Breech mech. | Elevation max (deg) | Traverse (deg) | Weight in action (kg) | Shell weight (kg) | Muzzle velocity (m/sec) | Range max (metres) |
|---|---|---|---|---|---|---|---|---|---|---|
| **Field artillery** | | | | | | | | | | |
| 76.2mm M00 | 1900 | 76.2 | 30 | SQF | 17 | 4 | 1020 | 6.5 | 588 | 6400 |
| 76.2mm M02 | 1902 | 76.2 | 30 | SQF | 17 | 5 | 1040 | 6.6 | 593 | 6600 |
| 76mm M02/30 | 1930 | 76.2 | 40 | SQF | 37 | 5 | 1350 | 6.4 | 680 | 13000 |
| 3in M13 | 1913 | 76.2 | 31 | SQF | 16 | 6 | 10030 | 6.6 | 510 | |
| 76mm Regimental Gun M27 | 1927 | 76.2 | 15 | SQF | 25 | 6 | 780 | 6.2 | 387 | 8550 |
| 76mm Divisional Gun M33 | 1933 | 76.2 | 50 | VSB | 43 | 5 | 1600 | 6.4 | 715 | 13600 |
| 76mm Divisional Gun F-22 | 1936 | 76.2 | 51 | VSB | 75 | 6 | 1620 | 6.4 | 706 | 13600 |
| 76mm Divisional Gun USV | 1939 | 76.2 | 42 | VSB | 45 | 60 | 1484 | 6.1 | 670 | 12185 |
| 76mm Gun M41 | 1941 | 76.2 | 43 | VSB | 18 | 54 | 1100 | 6.2 | 680 | |
| 76mm Gun ZIS-3 | 1942 | 76.2 | 42 | VSB/SA | 37 | 54 | 1116 | 6.2 | 680 | 13000 |
| 85mm Gun M02 | 1902 | 85 | | | 16 | 5 | 1930 | 6.5 | 555 | 6400 |
| 85mm Divisional Gun D-44 | 1944 | 85 | 55 | VSB/SA | 35 | 54 | 1725 | 9.5 | 793 | 15500 |
| 85mm Divisional Gun D-48 | 1945 | 85 | 55 | VSB/SA | 40 | 54 | 2100 | 9.5 | 793 | 16000 |
| 107mm Gun M10/30 | 1930 | 107 | 38 | SQF | 37 | 6 | 2380 | 17.2 | 670 | 16350 |
| 107mm Gun M40 | 1940 | 107 | 44 | HSB | 44 | 60 | 3955 | 17.2 | 720 | 17450 |
| 122mm Howitzer M04 | 1904 | 122 | 12 | SQF | 42 | 5 | 1225 | 21 | 292 | 6700 |
| 122mm Howitzer M10/30 | 1930 | 122 | 13 | SQF | 43 | 5 | 1465 | 21.7 | 364 | 8900 |
| 122mm Howitzer M38 | 1938 | 122 | 23 | SQF | 63 | 50 | 2250 | 21.7 | 500 | 11795 |

*Russian/Soviet Artillery Equipment (Cont)*

| Equipment | Date | Calibre (mm) | Barrel Length (cals) | Breech mech. | Elevation max (deg) | Traverse (deg) | Weight in action (kg) | Shell weight (kg) | Muzzle velocity (m/sec) | Range max (metres) |
|---|---|---|---|---|---|---|---|---|---|---|
| **Heavy artillery** | | | | | | | | | | |
| 122mm Corps Gun M31/37 (A-19) | 1938 | 122 | 45 | SQF | 65 | 58 | 7120 | 24.9 | 800 | 20800 |
| 122mm Gun M1955 (D-74) | 1955 | 122 | 45 | VSB | | | 5000 | 25.5 | 800 | 21000 |
| 122mm Gun-How D-30 | 1963 | 122 | 35 | VSB | 70 | 360 | 3175 | 24.9 | 716 | 15400 |
| 130mm Gun M46 | 1954 | 130 | 55 | HSB | 45 | 50 | 8450 | 33.5 | 930 | 27000 |
| 152mm Gun M10/30 | 1930 | 152 | 28 | IS | 37 | 5 | 6700 | 43.6 | 650 | 17100 |
| 152mm Gun M10/34 | 1934 | 152 | 29 | VSB | 45 | 58 | 7100 | 43.6 | 655 | 16200 |
| 152mm Gun BR-2 | 1935 | 152 | 45 | IS | 60 | 8 | 18350 | 48.9 | 880 | 26975 |
| 152mm How M-10 | 1938 | 152 | 25 | SQF | 65 | 50 | 4165 | 40 | 508 | 12400 |
| 152mm How D-1 | 1943 | 152 | 23 | SQF | 63 | 35 | 3600 | 40 | 510 | 12400 |
| 152mm Gun-How ML-20 | 1937 | 152 | 29 | IS | 65 | 58 | 7130 | 43.4 | 655 | 17280 |
| 152mm Gun-How D-20 | 1955 | 152 | 37 | VSB | 63 | 60 | 5900 | 48 | 670 | 18000 |
| 180mm Gun S-23 | 1955 | 180 | 45 | IS | 50 | 44 | 21450 | 84 | 790 | 30400 |
| 203mm How L-25 | 1931 | 203 | 25 | IS | 60 | 8 | 17700 | 98.5 | 606 | 18000 |
| 21cm Gun M39/40 | 1940 | 210 | 48 | IS | 50 | 22 | 43180 | 134.9 | 800 | 30430 |
| 30.5cm How BR-18 | 1940 | 305 | 22 | IS | 77 | 360 | 62110 | 330 | 530 | 16400 |

## RUSSIAN/SOVIET SELF-PROPELLED ARTILLERY

| Equipment | Date | Calibre (mm) | Chassis | Elevation max (deg) | Traverse on mount (deg) | Weight in action (kg) | Range, max (metres) | Ammunition carried (rds) |
|---|---|---|---|---|---|---|---|---|
| SU76 | 1943 | 76.2 | T70 tank | | | 11200 | | 60 |
| SU85 | 1943 | 85 | T34 tank | | | 29600 | | 48 |
| SU100 | 1944 | 100 | T34 tank | 17 | 16 | 31600 | 3000 (A/Tk) | 34 |
| ASU57 | 1957 | 57 | Special | 12 | 16 | 7400 | 1000 (A/Tk) | 40 |
| ASU85 | 1961 | 85 | PT76 tank | 15 | 12 | 14000 | 3000 (A/Tk) | 40 |
| JSU122 | 1943 | 122 | JS tank | | | 41200 | | 40 |
| JSU152 | 1943 | 152 | JS tank | | | 41800 | | 25 |
| M1973 (SO-152 Akatsiya) | 1973 | 152 | Special | 65 | 360 | 23000 | 18500 | 46 |
| M1974 (SO-122 Gvozdika) | 1974 | 122 | Special | 70 | 360 | 16000 | 15300 | 40 |
| M1975 (SO-203) | 1975 | 203 | Special | 60 | ? | 40000 | 30000 | ? |
| ZSU-57-2 | 1954 | 57 twin | Special | 85 | 360 | 28100 | 4000 (AA) | 316 |

# S

**ST CHAMOND** Common name of the Compagnie des Forges et Acieries de la Marine et d'Homecourt, located in St Chamond, France. This company began as an iron foundry in the early 19th century and then progressed to steel-making. By the 1880s it was producing high-quality armour plate and then began developing armour-piercing ammunition. It progressed to artillery, principally for naval use, but occasionally for land service, and provided the Mexican Army with a field gun in 1902 and the Belgian Army with a 120mm field howitzer in 1913. During the First World War it developed a number of field carriages to which naval guns could be fitted, to augment the French field artillery. St Chamond developed one of the first French tanks, and in 1918 produced one of the earliest self-propelled guns, a 24cm howitzer on a tracked chassis driven by electric motors. The current for the 24cm was delivered by flexible cable from a power tractor that accompanied the gun mounting. St Chamond abandoned the manufacture of ordnance after the First World War.

**SADDLE** The part of a gun carriage or mounting that supports the elevating mass and which traverses above the trail. Principally a British term; other nations use 'top carriage' (qv) interchangeably.

**SALVO** The firing of a number of guns simultaneously on order; as opposed to 'gunfire' in which the guns fire independently upon the order to fire being given.

**SCHLANKE EMMA** An Austrian 30.5cm howitzer developed by the Skoda company and first used against the fortress of Namur in 1914. Skoda, like Krupp, had begun developing coast defence howitzers of large calibre in the 1890s, and in the early 1900s began work on a heavy road-mobile howitzer. The first model was the 24cm 'Gretel', but this was soon superseded by a 30.5cm design that was extensively tested in 1911-12 and issued to the Austrian Army in 1913. The design consisted of a bed, buried in the ground, which supported a revolving top carriage into which the gun, in a ring cradle, fitted. The howitzer could be dismantled into three loads, towed by Austro-Daimler 100 horsepower tractors. It was assembled on site by a portable crane, and a well-trained detachment could have it in action in 25 minutes. It fired a 287kg (633lb) shell to 11 kilometres (6.8 miles) or a 380kg (838lb) shell to 9.6 kilometres (6 miles). These weapons went back to Austria after the Belgian and

A St. Chamond 28cm SP howitzer developed in 1918. The screw-jack prevented the recoil from damaging the suspension when the howitzer fired.

The Schneider self-propelled chassis mounting a 22cm model 1917 long-range gun.

French fortresses had fallen, escaped destruction after the war, and were emplaced as coast defence mortars around the Adriatic during the Second World War. The name 'Schlanke (Slender) Emma' was given by the Austrian gunners as a riposte to the German 'Dicke Bertha' 42cm howitzers that were used in the same siege actions.

**SCHNEIDER** Common name for the French firm Schneider-Creusot, manufacturers of artillery weapons, ammunition, armour plate and other munitions. Founded as an ironworks in 1836 by the brothers Adolphe and Eugéne Schneider, the company soon went into the manufacture of cannon and acquired a reputation for high quality and good design. It built a wide variety of guns for the French Army and also exported widely, pioneered armoured turrets and cupolas for fortress artillery,

and built tanks during and after the First World War. It became Schneider SA, a holding company, after the Second World War and has since been absorbed and amalgamated with other concerns to form one of the largest industrial organizations in the world. Its only armament function today is the manufacture of armour by a subsidiary, Creusot-Loire.

**SCHNEIDER BREECH MECHANISM** A form of breech mechanism for bag-charge guns developed by the Schneider company (see above). It used an interrupted screw and de Bange obturation, and was operated by a lever that lay along the top of the breech-screw carrier. Pulling the lever back and to the right caused the screw to rotate and the breech to swing open. Its most common application was on the French 155mm M1917 gun and howitzer, both of which were later adopted by the US Army.

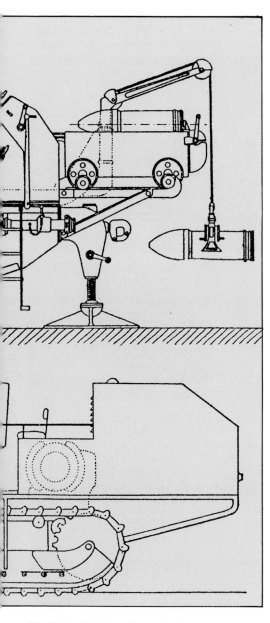

**The Schneider breech mechanism on a French 155mm howitzer M1917. The operating handle lies across the top, the hinge being on the right side.**

**SCOTT CARRIAGE** The name given to a variety of extempore gun carriages devised by Captain Percy Scott RN for mounting naval guns for use in land operations. These gun carriages were principally used during the South African War when a shortage of long-range guns led Scott to remove some 12pr and 4.7in guns from his ship HMS *Terrible* and place them on carriages made locally to his design, after which they were sent to Pretoria and other actions with naval crews.

**SCREW-GUN** The popular term for a 'jointed gun' in which the barrel is divided into breech and muzzle sections, the two being joined by a screwed collar. The object was to reduce the weight of the various component parts of the gun so that it could be carried on pack-saddles for operations in mountainous country. Widely used by British and Indian Army mountain artillery, the last model was the 3.7in Mountain Howitzer, which was declared obsolete in 1960.

**SELF-HOOPING** An alternative expression for 'Auto-frettage' (qv)

**SEMOVENTE** Italian self-propelled artillery equipment. The Semovente 90/53 consisted of a 90mm anti-aircraft gun mounted at the rear end of the hull of a modified M41 medium tank chassis. The gun was in a shielded limited-elevation mounting for use solely as an anti-tank gun, in imitation of the German adaptation of their 88mm anti-aircraft gun in the anti-tank role. About 30 were built in 1942-3 and saw limited use against the US and British forces during the invasion of Sicily. The 90mm gun fired a 10kg (22lb) armour-piercing shell at 840m/sec (2,760ft/sec) and could defeat 140mm (5½in) of armour at 500m (547 yards) range.

The Semovente 149 was a 149mm Model 35 gun mounted on a similar chassis to act as a self-propelled field weapon. It fired a 50kg (110lb) shell at 800m/sec (2,625ft/sec) to a range of 23.7 kilometres (14.8 miles). Only one was built, however, before Italy opted out of the war.

**SEXTON** Self-propelled gun equipment used by British and Commonwealth armies during and after the Second World War. After the US development of the M7 'Priest' SP howitzer (qv), the British Purchasing Commission asked for a variant model mounting the British 25pr gun. American policy was to build only those weapons used by the US forces, so the request was turned down. The project then went to Canada, where the Canadians had developed a 'Ram' battle tank based on the Amer-

ican Sherman. It was decided for various reasons not to use the 'Ram' as a combat tank, and after sufficient numbers had been made for training purposes the plant began building the chassis with an open-topped hull mounting the 25pr gun, which became known in British service as the 'Sexton'. It replaced the 'Bishop' (qv) and was a much better equipment insofar as the gun could reach its full 7.6 miles (12.25 kilometres) range in the new mounting. Sexton weighed 25.4 tons, carried 105 rounds of ammunition, and had a six-man crew. Introduced in 1943, a total of 2,250 was built and some examples remained in service with the South African and Portuguese armies until the late 1970s.

**SHIELD** A steel plate fitted on a gun carriage ahead of the axle to protect the gun detachment from bullets and shell splinters. It is pierced to permit the gun barrel to pass through and elevate, and also permit a direct sight to see forward. The shield may have flaps at top and bottom which can be unfolded once the gun is in action to increase the amount of protective area. It also generally carries clips and fittings to permit the carriage of accessories such as ropes, tool-boxes, rammer etc.

The shield was introduced with the French 75mm M1897 field gun, because at that time direct fire was still the general rule and the modern magazine rifle firing high velocity bullets, then entering service, meant that the gunners were at risk from infantry fire at ranges of one mile or more. The hardened steel shield gave adequate protection for the men because they could now be tightly grouped round the gun when it fired because of the adoption of a recoil mechanism. Hitherto, the gunners had to stay away from the gun to permit the entire gun and carriage to recoil, so there was no point having a shield. The shield was universally adopted for field guns in subsequent years and stayed in use until after the Second World War. It has now been abandoned, because it adds weight which is unwelcome in these days of helicopter-lifted guns, and as the gun now rarely fires in the direct role the risk of coming under infantry fire is negligble.

**Top left: Sexton, the Canadian-designed self-propelled 25pr gun used by British and Commonwealth armies.**

**Left: The shield of an Italian 105mm M56 pack howitzer – which would seem to have been a useful repository for stores as well as a means of protection.**

**SELF-PROPELLED** An artillery weapon carried on a motorized vehicular mounting; usually tracked, but can be wheeled.

Self-propelled (SP) artillery was first developed during World War One by the French, who produced a handful of SP mountings for heavy guns and howitzers. The idea was experimented with by other countries in the 1920s, and Britain produced a number of SP 18pr field equipments known as 'Birch Guns' (qv). Wide use of SP artillery came in World War Two when all armies adopted it. SP guns generally fall into two groups: those which are purely direct-fire weapons for accompanying infantry, known as 'assault guns'; and those which are standard indirect-fire support weapons that have been given mobility in order to improve their tactical flexibility. The former class is closer in its employment to tanks or tank destroyers, and frequently acts in these roles. The latter is used in the same way as wheeled equipments, having the advantage of quicker movement across country and being generally faster into and out of action.

**A typical modern SP gun: the US 175mm M107.**

**SEMI-AUTOMATIC BREECH** A breech mechanism that is automatically opened during the run-out of the gun after recoil so that when the gun comes to rest it is ready to be reloaded. It is generally found on guns where a high rate of fire is desirable (eg, anti-tank, anti-aircraft and anti-torpedo-boat guns), and nowadays on some 155mm howitzers to permit rapid bursts of fire.

The semi-automatic breech is most usually found with sliding block mechanisms, because these are relatively easy to actuate. A typical method is to have the breech operating lever arranged so that it will open the breech against the power of a spring. The extractors lock the block in the open position when the breech is open, and the lever can then be returned to its position of rest. When the cartridge is loaded, the rim of the cartridge case pushes the extractors forward, so releasing the breechblock, which is closed by the spring. The gun recoils on firing, and as it runs out a crank on the pivot shaft of the operating lever strikes a protrusion on the gun cradle which causes the shaft to revolve and open the breech until it is, again, held by the extractors ready for reloading. The act of opening the breech automatically ejects the spent cartridge case.

The semi-automatic breech of a 1937-vintage German 5cm anti-tank gun; the cylindrical casing holds the spring which closes the breech after loading.

Left: The open breech of a British 6in coast defence gun with the shot guide in place behind the chamber, ready for loading.

**SHOT GARLAND** A framework of wood or iron consisting of a number of squares into which round shot can be placed. The garland is placed on the ground and shot inserted into each square, after which shot can be piled up in layers as required. Without the garland the shot on the bottom row would not stay still to permit a pile to be built. It was used only with round shot for muzzle-loading guns.

**SHOT GUIDE** A curved tray attached to the lower edge of the breech ring and coupled to the breech mechanism so that as the breech is opened the tray moves into position behind the breech opening. It is used to support the projectile in line with the axis of the bore while the rammer is positioned and the projectile is rammed. Use of a shot guide prevents damage to the threads of the breech aperture and to the rear face of the 'A' tube.

**SIEGE ARTILLERY** Heavy artillery used for the bombardment of fortresses. Generally it consisted of mortars and howitzers as well as guns, so that explosive and incendiary projectiles could be fired into the besieged place while the guns battered the walls with solid shot to effect a breach. Guns and howitzers on special carriages were designed in the latter part of the 19th century, so that they could bombard the place while keeping their gunners under cover and protected from the danger of riposte from rifled small arms. Some forms of wheeled disappearing guns were developed through rarely employed for this task. With the proliferation of heavy artillery for general bombardment tasks in the First World War, the distinction disappeared and the term was never used again.

**SIEGFRIED** German super-heavy railway gun equipment consisting of the 38cm SK C/34 naval gun on a railway mounting. The design was begun by Krupp in 1938, based on the guns developed for the *Bismarck* class battleships, and by mid 1939 the design had been approved by the Army and eight equipments ordered. However, as three 40.6cm barrels were available the three mountings were fitted with the larger guns and two were sent to Poland to strengthen the defences of Danzig (see under 'Adolf' for further details). The three 40.6cm guns were removed from the mountings in 1941 for emplacement in coast defences in France, and the mountings were then returned to Krupp to have the original 38cm guns fitted. It was the summer of 1943 before this was done, and the remaining five of the original order were never proceeded with.

Siegfried had a 47-calibre barrel that was supported by girder bracing against droop. It fired a 495kg (1,091lb) long-range shell to 55.7 kilometres (35 miles) range, or an 800kg (1,764lb) shell to 42 kilometres (26 miles) range. The entire equipment weighed 294 tonnes in action, travelled on 32 wheels, was 31.3 metres (103 feet) long and was usually emplaced on a turntable in order to give it a sufficient field of fire.

**SIGHTS** Apparatus fitted to the gun for the purpose of directing fire accurately at the target. Two forms of sight may be met, direct-fire and indirect-fire sights.

Direct-fire sights are those used when the target is in view of the gun. In their simplest form they need be no more than a notch and a foresight bead placed on a suitable bar attached to the gun. The bar is linked to the gun by an elevating mechanism that allows it to be depressed to an angle proportional to the required range. The sight can be brought into alignment by elevating and traversing the gun, the gunner looking across it in a similar manner to the sights of a rifle. A more efficient system is to use a low-power telescope with cross-wires.

Indirect-fire sights become more involved as the target is no longer visible, and consist of a panoramic sight (qv) which can be directed at some fixed aiming point (qv). The required range can be set on a scale, which displaces a clinometer (qv) bubble, and elevating the gun to bring the bubble level will then apply the required quadrant elevation to the gun barrel.

Modern artillery, particularly self-propelled equipments, are now adopting electronic sights that will compensate for lack of level of the weapon platform, for drift and other functions, and that will also, if required, compute corrections and firing data, basing their computation on the position of the gun as determined by an inertial navigation system built into the vehicle. Towed equipments use sights that display digital information from the fire direction centre and can automatically incorporate gun corrections and muzzle velocity corrections, relieving the gunlayer from the responsibility of remembering these and incorporating them into the data ordered.

**The sights of the British 25pr gun. The cone is the range scale, with an adjustable reader, which is set to the propelling charge in use. Below it is the sight clinometer.**

Above: The prototype German 105mm light field howitzer of 1943, developed by Skoda. It used a cruciform mounting to permit all-round fire.

An experimental auto-loading 5cm anti-tank gun developed by Skoda in 1944.

**SINGLE-MOTION BREECH** The term applied to a screw breech mechanism in which all the motions of rotating the screw, withdrawing it from the breech and swinging it to one side are performed by a single movement of an operating handle. The Asbury Breech (qv) is a good example of the type. Compare with 'Continuous Motion' and 'Three-Motion' breeches.

**SITE** An American term for 'Angle of sight' (qv)

**SKODA** Founded in 1859 in Pilsen, then part of the Austrian Empire, the Skoda works rapidly became the largest steel plant in Central Europe and bore the same relationship to the Austro-Hungarian government as did Krupp's to the German government. It produced artillery, armour plate and ammunition as well as railway equipment, bridges and other heavy engineering products. Gun manufacture continued throughout the First and Second World Wars, the firm coming under the Czech regime after the foundation of that country in 1919. The Skoda works produced a wide variety of artillery, and was particularly noted for an excellent range of light mountain guns on the one hand and a series of super-heavy howitzers at the other end of the scale. The factory still operates, though it is now known as the 'V.I. Lenin Works', and manufactures armaments for the Warsaw Pact countries.

**SKYSWEEPER** American 75mm anti-aircraft gun M51, adopted for service in the early 1950s and withdrawn in the early 1970s.

Development of this weapon began in August 1944 when it was seen that the development of proximity fuzes would permit the design of a gun that did not require fuzes to be set before loading, thus permitting some form of automatic rapid loading. The calibre of 75mm was selected as being the smallest that would permit the shell to carry a proximity fuze (which was much more bulky than the standard time fuze of the period) and still hold a useful amount of explosive. The design was intended to provide a totally autonomous weapon, having its own radar, optical director and fire control computer on the mounting. An automatic loading system, using two revolving drums behind the breech feeding to a central rammer, allowed a rate of fire of 45 rounds per minute. The gun was built at Watervliet Arsenal and was ready by January 1945, but the complex mounting was not ready until the summer of that year. However, test-firing of the gun showed that it only achieved a muzzle velocity of 700m/sec (2,297ft/sec), which was nowhere near the figure needed for an efficient anti-aircraft gun, and the design was refused. The designers went back to try again, with a target of 915m/sec (3,002ft/sec), but in the end the gun only managed 855m/sec (2,805ft/sec) velocity. This was accepted, but the development of the predictor/ radar combination again delayed production and it was not until 1951 that the final design was approved.

**A battery of American 75mm Skysweeper anti-aircraft guns. The corporal is operating the optical tracker. The radar dish is visible under the barrel.**

**SLEIGH** A component, usually a steel casting, that is interposed between the gun barrel and the cradle in some designs of gun. There are two principal applications of the sleigh:

1, In pack howitzers, which need to be dismantled and re-assembled quickly, the sleigh is permanently attached to the cradle by the recoil mechanism rods, and the gun barrel drops into the sleigh and is locked by some quick-acting device. This saves having to attach the gun to the recoil system by unscrewing and screwing-up nuts.

2, In high-angle howitzers the sleigh slides in the recoil guides in the elevating cradle, but the barrel is attached to the sleigh so that it can pivot independently of whatever elevation the cradle might have. Thus, the gunlayer can keep the cradle at the correct elevation, but the barrel can be unlocked and swung down to the horizontal for loading, and returned and locked to the cradle for firing. In this application it performs the same task as a quick-loading gear (qv). An example of this is the US 12in Coast Mortar M1912.

**SLIDE-BOX** The metal holder screwed to the end of the vent of a screw breech and on which the firing lock is carried.

**SMITH GUN** The Smith Gun was a unique weapon developed as a 'private venture' in 1940 and offered to the British Army as an emergency gun for issue to the Home Guard as an anti-invasion defence. It was a 3in (76.2mm) smoothbore barrel mounted in a light carriage that consisted of little more than two wheels and an axle; the gun passed through a hole in the axle and the recoil was absorbed by a number of rubber rings round the barrel that were squeezed against the axle when the gun fired. The wheels were steel discs, dished so that one wheel was concave and the other convex; to place the gun in action it was tipped over on to the concave wheel, which then formed a platform round which the gun could be traversed, the axle forming the pivot. Traverse was free, the gunner merely pushing down on a lever behind the gun and clamping it when the correct elevation was reached. A similarly-wheeled limber was provided to carry ammunition, and both items could be towed behind the average private car. The projectile was a cylinder of cast iron containing a charge of explosive and a fuze, propelled by a small charge of smokeless powder fired by a .38 pistol blank cartridge.

The name 'Smith Gun' came from the inventor, who was the chief engineer of the Trianco Engineering Company. At first the gun was refused by the Ordnance Board, but it appears to have got into Home Guard service by some back-door route and was on issue by June 1941. In addition to the large numbers built and issued to the Home Guard, a few were also issued to Army and Royal Air Force units guarding airfields. None saw action, and they were declared obsolete in 1945.

The breech of the Smith gun opened upwards on two arms. The ammunition rack to the right holds one cartridge in the topmost container.

The breech of a British 6in gun, showing the slide box supporting the firing lock and enclosing the rear end of the gun vent.

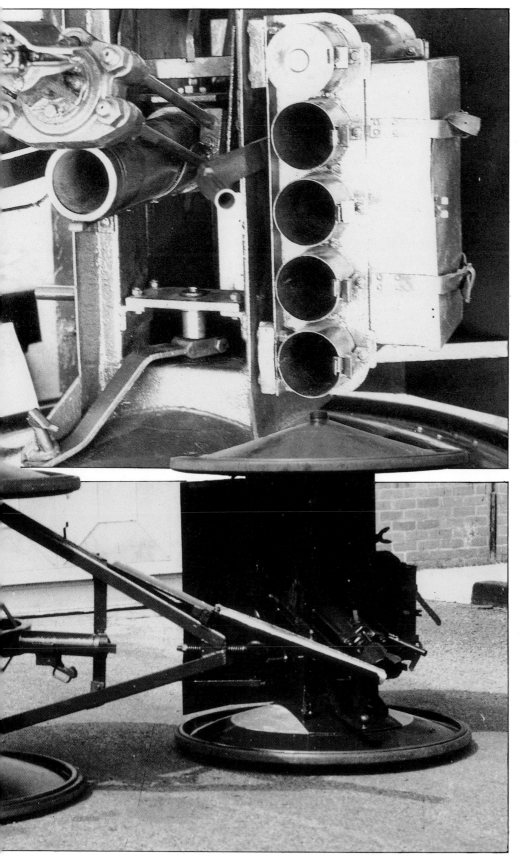

**SNIFTING VALVE** A pressure relief valve in the recuperator cylinder of a gun recoil system which releases any build-up of air due to leaking packings.

**SOFT RECOIL** An American term introduced in the 1960s and meaning the same as 'differential recoil' (qv).

**SOLE PLATE** A steel plate, usually fitted with a toothed rack, which lies on the ground beneath the trail end of a large gun or howitzer whose carriage is supported by a platform with pivot. A gear-wheel, actuated by a hand crank and mounted on the gun trail, engages with the toothed rack by operating the crank, so traversing the gun mounting.

**SOUND CANNON** A weapon developed in Germany during the Second World War which was designed to cause casualties and damage by means of sound waves of great intensity. Invented by a man called Wallauschreck, an Austrian, it consisted of a parabolic reflector 3.2 metres (10.5 feet) in diameter, with an attachment extending to the rear of the vertex of the parabola. This attachment consisted of the firing chamber, the length of which was the wavelength of the developed sound. The firing chamber had two nozzles at its rear through which oxygen and methane were injected into the chamber, where the mixture was then ignited. Explosions were produced at a rate between 800 and 1,500 times a second, developing a noise which was then focussed and directed by the parabolic reflector. It was claimed that a range of 60 metres (66 yards) the weapon could kill a man, and at 300 metres (328 yards) seriously disable him for an appreciable length of time. This may have been true, but it was a totally unpractical weapon which, apart from an experimental model, was never adopted.

**SOUND RANGING** A method of locating the position of hostile artillery by burying a line of microphones in the ground and connecting them to a recording instrument. The sound wave from a gun firing will pass over these microphones at slightly different times due to the radial dispersal of the wave, and by mathematical analysis it is possible to determine the bearing to the source of the sound from any pair of microphones. Thus with, say, six microphones spread over a five-mile front it becomes possible to generate a number of lines that will meet at the calculated position of the gun. The system is somewhat susceptible to wind and other conditions, but when working properly can deduce positions to an accuracy of better than 50 yards (46 metres). The system can also be used to range gunfire on to the distant target by comparing the sound record of the target gun with those of the shells; when the two patterns agree, the shell must be landing on the gun.

**A Smith gun turned on its side and ready for action with its ammunition trailer.**

**SPADE** Blade beneath the gun trail to bite into the ground and resist any rearward movement of the carriage due to recoil forces. It is usually at the trail end, but in some heavy guns has been fitted to the mounting body or to the trail close to the mounting body.

**SPANNING TRAY** A curved tray that reaches to the breech of the gun from some other fixture, such as a loading table, and spans the gap behind the gun where there is no space for the loaders to stand; eg, where there is a deep recoil pit immediately behind the breech, the spanning tray allows the shell and cartridge to be rammed across it into the breech.

**Right: A rear view of a German 75mm anti-tank gun, showing the spades at the end of the trail legs.**

**Left: Loading an American 240mm howitzer. The cartridge has just been pushed across the spanning tray into the chamber.**

**SOUTH AFRICAN ARTILLERY** The South African Defence Force continued to use British artillery after the country's separation from Britain, and the 25pr and 5.5in guns are still in active use. In addition, the SADF uses the universal 40mm Bofors light AA gun and various Oerlikon 20mm and 35mm automatic AA guns.

In 1975 the South African artillery, finding that its equipment was being outranged by Soviet weapons used by Angolan forces, set a requirement for a new long-range gun, which has since been filled by the Armscor G-5 155mm howitzer. This is an auxiliary-propelled weapon capable of firing to a range of 30 kilometres (19 miles) with conventional projectiles and to 39 kilometres (24 miles) using long range 'base bleed' shells. The same howitzer has since been fitted to the specially designed six-wheeled armoured chassis of the G-6 wheeled SP gun. This has advantages over tracked vehicles in economy and ease of maintenance, but has not yet been put into production.

**The South African G5 155mm howitzer**

## SOUTH AFRICAN ARTILLERY EQUIPMENT

| Equipment | Date | Calibre (mm) | Barrel length (cals) | Breech mech. | Elevation max (deg) | Traverse (deg) | Weight in action (kg) | Shell weight (kg) | Muzzle velocity (m/sec) | Range, max (metres) |
|---|---|---|---|---|---|---|---|---|---|---|
| **General support artillery** G-5 | 1983 | 155 | 45 | IS/SA | 75 | 82 | 13500 | 46 | 897 | 30000 |

| **Self-propelled artillery Equipment** | Date | Calibre (mm) | Chassis | Elevation max (deg) | Traverse on mount (deg) | Weight in action (kg) | Range, max (metres) | Ammunition carried (rds) |
|---|---|---|---|---|---|---|---|---|
| G-6 | 1982 | 155 | 6 wheel | 75 | 80 | 36500 | 40000 | 47 |

**SPIKING** The process of rendering a gun incapable of use to an enemy if captured. In the days of muzzle-loading smoothbores this was simply achieved by driving an iron spike into the vent, hammering it well in so that it flowed into the irregularities in the vent and could not be removed. The only way such a gun could be used was by drilling a new vent, but this could not be done quickly in the field and therefore the gun could not be used after capture. The 'spring spike' was invented in the 19th century; it had a sprung barb that went through the end of the vent into the gun chamber and there sprang open, so preventing the spike being withdrawn. Provided the man inserting the spike remembered to put it in the right way round, the spring could be compressed by a skilful use of a rammer inserted down the bore and the spike could be withdrawn when the gun was recaptured. The spring spike was used when the battle was ebbing and flowing and there was a good chance that the enemy would be counterattacked and the guns recaptured before the enemy could put them to use.

With the adoption of breech-loading, spiking was no longer possible, though the term has remained in common use. The standard technique to disable a breech-loading gun is to insert a shell nose-first into the gun's muzzle, then load a shell and cartridge in the normal manner in the breech, attach a long line to the firing mechanism, take cover and fire the gun. The two shells meet and detonate, and the gun is, theoretically, destroyed. What happens in practice is that the end of the barrel is blown off, more or less cleanly. It has been known for an enemy to put the gun to use as soon as it has been captured, ignoring the loss of range due to the shortened barrel. The most effective method is to load a fuzeless shell backwards into the chamber, leaving the open fuze-well and exposed explosive pointing to the rear. A cartridge is then loaded and fired. The propellant flame detonates the shell in the chamber and the usual result is to blow off the breech end and thoroughly wreck the recoil mechanism.

**SPLIT-TRAIL CARRIAGE** A type of carriage used with guns and howitzers, in which the trail consists of two legs, hinged to the axle-tree, and capable of being swung open to an angle of about 40°. The gun thus has greater stability, though some form of articulation (qv) must be used to compensate for uneven ground. It allows the gun barrel to have a much greater arc of traverse than is usual with a pole or box trail because the line of the thrust of recoil is resisted by the widely-set trails. On the other hand, a powerful gun firing at extreme traverse on a light carriage can often make the opposite trail leg lift into the air and give the appearance of instability. A further advantage is that the gun detachment can work within the trail legs, close to the breech, which is often more convenient than leaning across a wide trail to load and operate the breech. Its sole disadvantage is that the legs have to be longer than a one-piece trail; this is because the stability, or resistance to overturning when fired, is a function of the distance from the axle to the tip of the trail, and to obtain this distance when the trail legs are open means that the trails when closed for travelling are longer than normal which, of course, also means they are heavier.

The split-trail carriage was invented by Colonel Deport (qv) of France and was first applied to the Deport M1912 75mm field gun, designed for the Italian Army. Deport adopted the split-trail for this gun because the Italians wanted a weapon that could act as a normal field gun but could also reach a high elevation (65°) and act as an anti-balloon gun, and the split trail allowed the extra elevation.

**Below: An unusual split-trail carriage carrying an American 90mm anti-tank gun, 1945.**

**SPRING SPADE** A system of recoil control used on some field guns in the 1890-1900 period. It was generally adopted as a method of controlling recoil on guns that had been built in the period before on-carriage recoil systems were invented in an endeavour to bring the gun up-to-date. It consisted of a steel arm, hinged to the underside of the trail, and fitted with a spade at the bottom end. The arm was linked to the trail by a powerful spring. The spring spade was released when the gun was in action so that the spade rested on the ground and the arm sloped backwards. The spade end would dig into the ground when the gun fired; the gun would recoil over it, and as it did so the spring spade arm would be pulled into the vertical against the power of the spring. This would absorb some of the recoil and, once recoil had ceased, the power of the

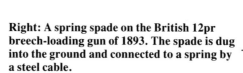

**Right: A spring spade on the British 12pr breech-loading gun of 1893. The spade is dug into the ground and connected to a spring by a steel cable.**

spring, pulling on the arm, would help to return the gun carriage to its original position. It was not a particularly effective device and did not last long.

**SQUEEZE BORE** A ballistic system in which the gun barrel is parallel for much of its length, then tapers rapidly to a smaller calibre and thereafter remains at the smaller calibre; it is a variant of 'Taper Bore' (qv), which is easier to manufacture and has the same effect. The only example of this system to achieve military service was the German 7.5cm PaK41 anti-tank gun, in which the conventional 75mm barrel was 295cm (9.7ft) long and had a squeeze bore section 95cm (37in) long attached by a threaded collar. As the projectile entered the squeeze it passed through a section tapered at 1 in 20 for 27cm, then at 1 in 12 for 17cm and finally into a parallel 55mm bore for the remainder of its travel. The whole of the squeeze unit was unrifled and could be quickly changed in the field when it wore out. The projectile for use in this weapon had soft steel skirts that collapsed during the squeeze process, and the emergent velocity was 1,125m/sec (3,691ft/sec).

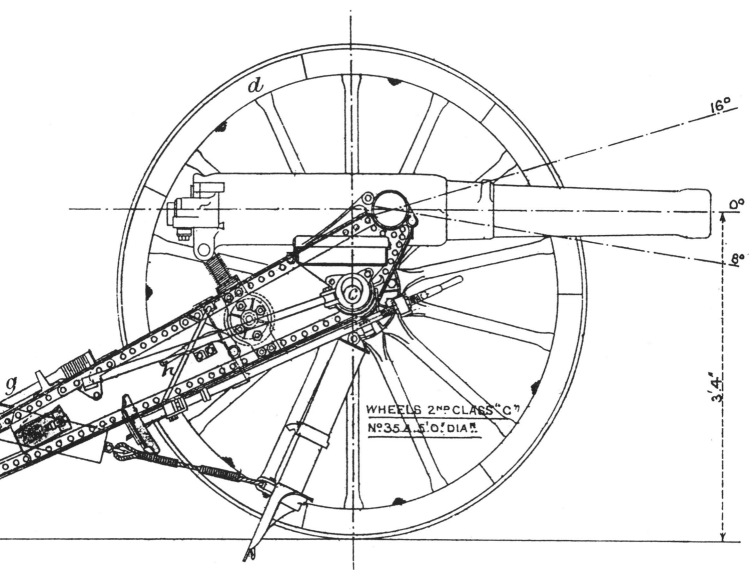

WHEELS 2ND CLASS "C"
No 354 5'0" DIA.

16°
0°
8°
3'4"

d
g
h

**S.R.C.** The abbreviation for Space Research Corporation; A Canadian company that developed a design of 45-calibre 155mm howitzer in the 1970s together with an Extended Range Full Bore (ERFB) projectile that carried aerodynamic stub wings on its forward portion. The design subsequently appeared in Austria (as the Noricum G-45); Belgium (as the SRC International GC45); and South Africa (as the G-5). Elements of the design can also be seen in other 155mm designs produced experimentally in China, Taiwan and other countries. The Canadian company no longer exists, but the development has been continued by SRC International SA of Belgium.

**STEEL CHOKE** A defect that sometimes appears in built-up guns, and takes the form of a ring round the interior of the bore which is below calibre and could interfere with the passage of the shell. It is caused by the internal hoops or tubes of the gun moving and thus causing the joints to exert pressure on the 'A' tube (qv).

**STOCKETT BREECH** A type of screw-breech used with heavy American guns developed between 1895 and 1920, particularly coast defence

guns. It was a form of continuous motion (qv) breech mechanism in which rotation of a crank alongside the breech first rotated the screw to unlock it; then, by engaging a gear with a 'translat-

ing rack' cut into the side of the screw, withdrew the screw on to a tray, and finally swung the tray to carry the screw clear of the breech opening.

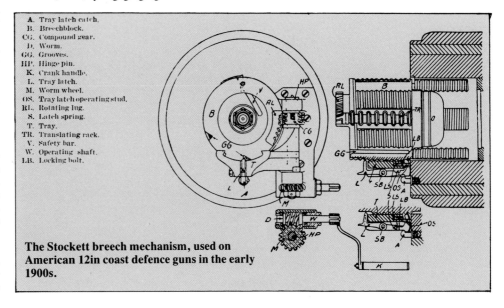

A. Tray latch catch.
B. Breechblock.
CG. Compound gear.
D. Worm.
GG. Grooves.
HP. Hinge pin.
K. Crank handle.
L. Tray latch.
M. Worm wheel.
OS. Tray latch operating stud.
RL. Rotating lug.
S. Latch spring.
T. Tray.
TR. Translating rack.
V. Safety bar.
W. Operating shaft.
LB. Locking bolt.

**The Stockett breech mechanism, used on American 12in coast defence guns in the early 1900s.**

**One method of sub-calibre training: a Bren machine gun, set to fire single shots, aligned with the barrel of a 25pr gun. The trigger is connected to the gun's firing lever. This system was used for training in anti-tank shooting.**

**STONK** A British artillery term for a 'Standard Linear Concentration'; ie, a line of shells falling across a front of 525 yards (480 metres), this figure being derived from the effective width of the fire of two field batteries side by side, with the third battery superimposed across the two. Devised during the Second World War as a fast method of delivering supporting fire in front of infantry, the production of the 'stonk' was reduced to a quick drill by using graphical templates on a plotting board. Unfortunately, the name had a certain ring about it and was rapidly adopted by other arms, though without knowing the precise meaning, so that 'to stonk' became a common expression for any form of artillery fire against a small target. The term, in its proper form, is now obsolete, but lingers on as a slang term.

**STRAUSSLER** Hungarian-born Nicholas Straussler became an engineer with the Alvis Car Company in Britain in the 1930s and was responsible for the design of a number of armoured cars sold in the export market. During the Second World War

he invented the 'Duplex Drive' method of enabling tanks to swim ashore from landing craft. His sole contribution to artillery design was to develop an auxiliary-propelled version of the 17pr anti-tank gun. The standard axle was widened and a small truck engine and transmission fitted, and small wheels were added to the trail ends. This enabled the gun to be driven into position without the need for a tractor. It was tested and found successful, but the idea was abandoned because of the extra work required in digging a gunpit to conceal the much-enlarged carriage. Although not formally adopted, this appears to be the first application of auxiliary propulsion (qv) to a field gun carriage.

**SUB-CALIBRE** Smaller than the calibre of the gun. Applied to artillery equipments, the term refers to adaptors and attachments that allow the firing of ammunition of smaller calibre than that standard for the gun. This is generally cheaper, ranges to a shorter distance, and is therefore of value for training and practice purposes. Sub-calibre equipment may be in the form of 'aiming rifles'

(qv); or complete small gun equipments fixed to the barrel of the parent gun by clamps and fired, by means of some linkage, by the parent gun's firing lever or lanyard; or small-calibre barrels held in calibre-sized rings and inserted inside the barrel of the parent gun. The calibre of these devices depends upon the size of the parent gun and the effect desired; for example, a small anti-tank gun could have a barrel insert of the standard military rifle calibre and fire ordinary rifle cartridges at short range for target practice. A 15in coast defence gun could have a 75mm gun strapped to its barrel firing high explosive shell to several thousand yards for practice in both shooting and observation of fire.

The term is also occasionally used to describe a projectile that is smaller in diameter than the calibre of the gun but fits the bore by being carried in a 'sabot' or shoe discarded at the muzzle, leaving the projectile to fly to the target. This system gives higher velocity and greater range because the projectile and sabot are lighter than a full-calibre projectile. It is, however, more usual to describe this type of ammunition as 'discarding sabot'.

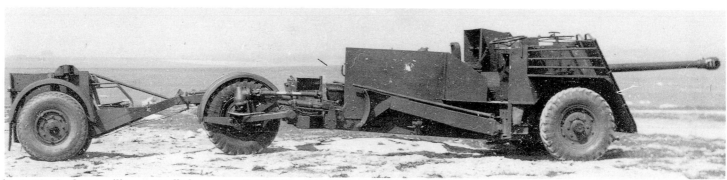

**Probably the first auxiliary-propelled gun, the British 17pr anti-tank gun modified by Nicolas Straussler in 1944.**

**SWEDISH ARTILLERY** Sweden's artillery is virtually synonymous with 'Bofors' (qv), yet despite that company's manufacture of guns since the 1880s the Swedish Army bought many foreign weapons until the First World War. The standard field piece was a Krupp 75mm gun, supported by a Bofors 105mm howitzer, and it was not until the 1930s that the equipment was modernized, largely by fitting the old barrels on to new carriages capable of more elevation and by providing more powerful ammunition. Faced with a major European war on its doorstep, in 1940 Sweden began modernizing its army at high speed and adopted a new range of artillery made by Bofors. More new Bofors designs have been introduced since then, including the innovative 'Bandkanon' self-propelled gun. Sweden is one of the few countries to maintain its coast artillery defences in good order, and in 1985 a number of improved 155mm guns were taken into service for mobile coast defence. There are also a number of turreted permanent coast defence batteries.

A 105mm field howitzer commercially developed by Bofors in the 1930s but not adopted by Sweden.

## SWEDISH ARTILLERY EQUIPMENT

| Equipment | Date | Calibre (mm) | Barrel length (cals) | Breech mech. | Elevation max (deg) | Traverse (deg) | Weight in action (kg) | Shell weight (kg) | Muzzle velocity (m/sec) | Range, max (metres) |
|---|---|---|---|---|---|---|---|---|---|---|
| **Anti-tank artillery** | | | | | | | | | | |
| 37mm Gun L/45 | 1934 | 37 | 45 | VSB/SA | 25 | 50 | 335 | 0.7 | 800 | 4500 |
| 57mm Gun L/50 | 1943 | 57 | 50 | VSB | 20 | 56 | 925 | 2.5 | 850 | 10500 |
| 75mm Gun L/54 | 1943 | 75 | 54 | VSB/SA | 20 | 30 | 2150 | 6.5 | 815 | 9800 |
| **Anti-aircraft artillery** | | | | | | | | | | |
| 40mm M33 Bofors | 1933 | 40 | 60 | VSB/A | 90 | 360 | 1730 | 0.9 | 900 | 4600 ceiling |
| 75mm L/52 Bofors | 1936 | 75 | 52 | HSB/SA | 85 | 360 | 4000 | 6.3 | 840 | 12000 ceiling |
| 80mm L/50 Bofors | 1936 | 80 | 50 | HSB/SA | 85 | 360 | 3500 | 8 | 750 | 15000 ceiling |
| **Mountain artillery** | | | | | | | | | | |
| 75mm Gun L/22 | | 75 | 23 | HSB/SA | 50 | 6 | 790 | 6.5 | 470 | 10000 |
| 9cm How Bofors | 1935 | 90 | 17 | HSB/SA | 50 | 6 | 790 | 9 | 350 | 7650 |
| **Field artillery** | | | | | | | | | | |
| 75mm Gun M1902 | 1902 | 75 | 30 | HSB | 16 | 7 | 975 | 6.5 | 500 | 7000 |
| 75mm Gun M02/33 | 1934 | 75 | 30 | HSB | 42 | 50 | 1350 | 6.5 | 540 | 11000 |
| 75mm Gun M40 | 1940 | 75 | 40 | HSB | 45 | 50 | 1500 | 6.3 | 700 | 14000 |
| 75mm Light Gun | 1935 | 75 | 23 | HSB/SA | 50 | 6 | 920 | 6.5 | 500 | 10200 |
| 105mm Gun M28 | 1927 | 105 | 42 | SQF | 45 | 60 | 3300 | 16 | 725 | 16200 |
| 105mm Gun L/42 | 1934 | 105 | 42 | HSB | 42 | 60 | 3750 | 15.3 | 785 | 17300 |
| 105mm Gun M37 | 1937 | 105 | 50 | HSB | 45 | 60 | 4500 | 15.3 | 800 | 16300 |
| 105mm Howitzer M1910 | 1912 | 105 | 16 | HSB | 43 | 4 | 1225 | 14 | 305 | 6200 |
| 105mm Howitzer M40 | 1940 | 105 | 22 | HSB | 45 | 50 | 1840 | 15.4 | 460 | 10000 |
| 105mm Howitzer Bofors 4140 | 1955 | 105 | 32 | VSB/SA | 60 | 360 | 2800 | 15.3 | 640 | 15600 |

Left: The Bofors 8cm field Anti-aircraft gun of 1930. The tubular mass at the breech end is actually the carriage pedestal, folded for transport.

Right: The latest of the Bofors line, the 57mm 'Trinity' automatic gun undergoing trials.

Below: The Bofors 9cm field gun, about 1935, at full recoil.

*Swedish Artillery Equipment (Cont)*

| Equipment | Date | Calibre (mm) | Barrel Length (cals) | Breech mech. | Elevation max (deg) | Traverse (deg) | Weight in action (kg) | Shell weight (kg) | Muzzle velocity (m/sec) | Range max (metres) |
|---|---|---|---|---|---|---|---|---|---|---|
| **Heavy artillery** | | | | | | | | | | |
| 15cm Howitzer M23 | 1923 | 149 | 17 | IS | 41 | 6 | 4320 | 41 | 500 | 12500 |
| 15cm Howitzer M37 | 1937 | 149 | 22 | HSB | 45 | 60 | 4300 | 41 | 475 | 12000 |
| 15cm Howitzer M39 | 1939 | 149 | 24 | HSB | 60 | 45 | 5700 | 42 | 580 | 14600 |
| 155mm Bofors How FH77A | 1975 | 155 | 38 | VSB/SA | 50 | 50 | 11500 | 42.4 | 774 | 22000 |
| 155mm Bofors How FH77B | 1982 | 155 | 39 | IS | 70 | 60 | 11900 | 42.4 | 827 | 24000 |
| 152mm Gun M37 | 1937 | 152 | 43 | IS | 45 | 60 | 13800 | 46 | 825 | 22500 |
| 152mm Gun M39 | 1939 | 152 | 43 | HSB | 45 | 60 | 14800 | 46 | 825 | 22500 |
| | | | | | | | | | | |
| **Coast artillery** | | | | | | | | | | |
| 120mm Static Gun 'Ersta' | 1985 | 120 | 62 | VSB | 55 | 360 | | 24 | 880 | 27000 |
| 120mm Mobile Gun 'Karin' | 1982 | 120 | 55 | VSB | 45 | 60 | 12000 | 24 | 840 | 20000 |
| 75mm Bofors L/60 | 1966 | 105 | | | 20 | 360 | | 5.7 | 850 | 12000 |

## SWEDISH SELF-PROPELLED ARTILLERY

| Equipment | Date | Calibre (mm) | Chassis | Elevation max (deg) | Traverse on mount (deg) | Weight in action (kg) | Range, max (metres) | Ammunition carried (rds) |
|---|---|---|---|---|---|---|---|---|
| Bandkanon 1A | 1965 | 155 | Special | 40 | 30 | 53000 | 24600 | 14 |

**SWISS ARTILLERY** The Swiss Army purchased its artillery from foreign sources until the 1940s, though some of the designs were locally manufactured under licence from about 1935 onwards. The originator of this equipment was generally Krupp, but later Bofors when Krupp weapons became unavailable during the Second World War. A number of local designs, developed by the Federal Gun Factory, Thun, have been taken into service since about 1948, together with some American 155mm SP howitzers.

Switzerland has a very large, but undisclosed, number of fortress guns of calibres from 75mm upwards; no details of any of these weapons, upon which the primary defence of Switzerland relies, have ever been made public.

**Above: The Oerlikon 35mm twin-barrel anti-aircraft gun on test in 1984. This gun is controlled by the one man seen here, ammunition feed being completely automatic.**

A self-contained self-propelled anti-aircraft gun, the Oerlikon 'Escorter' has two 35mm guns, electro-optical sights and radar on a wheeled mounting.

| Equipment | Date | Calibre (mm) | Barrel length (cals) | Breech mech. | Elevation max (deg) | Traverse (deg) | Weight in action (kg) | Shell weight (kg) | Muzzle velocity (m/sec) | Range, max (metres) |
|---|---|---|---|---|---|---|---|---|---|---|
| **SWISS ARTILLERY EQUIPMENT** | | | | | | | | | | |
| **Anti-tank artillery** | | | | | | | | | | |
| 47mm IK35 | 1935 | 47 | 31 | HSB | 55 | 48 | 254 | 1.8 | 567 | 8200 |
| 90mm M50 | 1950 | 90 | 32 | VSB/SA | 32 | 66 | 556 | 1.9 | 600 | 3000 |
| 90mm M57 | 1957 | 90 | 33 | VSB/SA | 23 | 70 | 570 | 2.7 | 600 | 3000 |
| **Mountain artillery** | | | | | | | | | | |
| 75mm Gun M33 | 1933 | 75 | 22 | HSB | 50 | 6 | 790 | 6.5 | 500 | 10500 |
| **Field artillery** | | | | | | | | | | |
| 75mm Gun M 03/22 | 1923 | 75 | 30 | HSB | 25 | 8 | 1096 | 6.4 | 485 | 10000 |
| 105mm Howitzer M46 | 1946 | 105 | 22 | HSB | 67 | 72 | 1840 | 15.1 | 490 | 10000 |
| 105mm Gun M35 | 1935 | 105 | | HSB | 45 | 60 | 4245 | 15.1 | 800 | 17500 |
| **Heavy artillery** | | | | | | | | | | |
| 15cm Heavy How M42 | 1942 | 149 | 28 | | 65 | 45 | 6500 | 42 | 580 | 15000 |

# T

**TAIL ROD** An indicator rod used with floating piston type recoil mechanisms in which a piston, not connected to anything, is interposed between the oil of the buffer and the compressed gas of the recuperator. The tail rod protrudes from the recoil system block and gives an indication of the balance of oil and air in the system. It is sometimes colloquially known as the 'tell-tale rod', but the correct name derives from its location as a 'tail' on the floating piston.

**TAMPION** A round plug that fits into the gun muzzle in order to exclude rain and dirt when the gun is not in use. It is rarely encountered today except on naval guns and has been superseded in army artillery by simple canvas waterproof covers.

**TANGENT ELEVATION** The angle between the line of sight from gun to target and the axis of the bore when the gun is laid.

**TANK DESTROYER** A form of self-propelled gun carrying an anti-tank gun and specifically intended for the destruction of tanks. Tank destroyers generally resemble tanks insofar as they mount their gun in a turret, whereas assault guns (qv) mount theirs in restricted-arc fixed mountings.

The term originated in the USA in 1941 with the formation of Tank Destroyer Battalions which, after various expedients had been tried and abandoned, were equipped with the M10 carrying a 76mm gun. This, in British service, was replaced by a 17pr gun, and in American service was superseded by a 90mm gun.

The tactical employment of tank destroyers relies upon their mobility to place the weapon in a suitable ambush, kill a tank, and move rapidly to another selected ambush site. Unfortunately, the vehicles' resemblance to a tank often gave their commanders the impression that they could charge about the battlefield mixing it with tanks on equal terms. They couldn't, and suffered accordingly.

**TAPER BORE** A ballistic system in which the bore of the gun tapers evenly from breech to muzzle so that the calibre is reduced. The ammunition needs to be specially designed so that the projectile will accommodate this reduction in calibre. This is generally achieved by soft metal skirts that deform and are pressed into recesses in the shell body so that when the shell emerges from the muzzle it is smooth and without protruding excrescences that could set up drag. The object is to increase velocity; this occurs because the base area of the projectile gradually decreases, while the pressure generated by the exploding charge remains more or less the same, as a result of which the unit pressure per square centimetre on the shell base is increased, so increasing the driving impulse and hence the velocity

The taper bore was pursued by Gerlich (qv) as a method of obtaining high velocity and hence a flat trajectory and greater accuracy in sporting rifles. It was adopted by both Rheinmetall and Krupp in Germany during the 1930s as a method of obtaining high velocity from anti-tank guns to improve their penetrative performance. Experiments were also conducted with taper bore barrels in larger calibres in order to improve the range of heavier artillery, but none of these projects came to fruition.

The principal drawback with the taper bore system is the difficulty of manufacturing a tapering bore, and for this reason the squeeze bore (qv) became a more favoured method. The complexity of the ammunition is another drawback, but it is one that was thought to be worthwhile in view of the considerable improvement in armour-piercing capability. Exact comparisons are apt to mislead, but the German 7.5cm PaK40, a conventional gun, could defeat 154mm (6in) of armour at 500 metres (547 yards) range, while the 7.5cm PaK41, a squeeze bore gun, could defeat 209mm (8in) at the same range.

The Germans' use of taper and squeeze bore guns came to an end in 1942 when what little tungsten (used for the penetrating projectiles) they had was reserved for machine tool production. The British development of discarding sabot ammunition in 1943-4 produced the same sort of results from conventional barrels and rendered the taper-bore obsolescent.

**The German Jagdpanther, powerfully armed with an 88mm gun and thickly armoured, was the archetypal tank destroyer of World War Two.**

**THEODOR** German railway artillery equipment using a 24cm gun on a double-bogie rail mounting. Developed in 1936 by Krupp, it used the 24cm SKL/40 ex-naval gun mounted in a ring cradle and carried in a fairly simple girder box mounting on two eight-wheel bogies. Three were built and issued for service in 1937. The gun was capable of 45° elevation and was provided with a portable turntable that could be rapidly installed at the end of any convenient railway spur to give an all-round field of fire. The shell weighed 148.5kg (327lb), had a muzzle velocity of 810 m/sec (2657ft/sec), and gave a maximum range of 26.75km (16.6 miles). The entire equipment weighed 95 tonnes and was 18.45m (60ft) long.

**THREE-MOTION BREECH** A screw breech mechanism in which the three funcions of rotating the screw, withdrawing it from the chamber and swinging it clear of the mouth of the chamber to permit loading were carried out in three separate steps. This was the earliest type of screw breech mechanism, dating from the 1880s, and was superseded by the continuous-motion breech (qv). See also 'Single-motion breech'.

**Above: A three-motion breech for the British 9.2in gun of 1889.** The central handle, assisted by the ratchet lever on the left, rotates the screw; the two grips allowed the screw to be withdrawn on to the carrier, and the crank handle then rotated the carrier to open the breech.

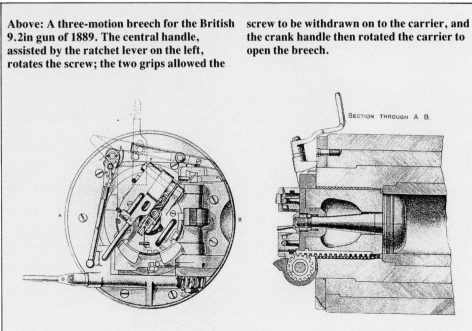

Above: An American 12in disappearing gun with a time/range board on the emplacement wall.

**THRUST COLLARS** Raised bands or collars round the body of a gun which engage with lugs on the recoil mechanism block in the cradle and locate and secure the gun so that the recoil force is transferred to the recoil system.

**TIME/RANGE BOARD** A system of fire control used with American coast defence artillery in the early years of the 20th century. It was a blackboard on the wall of the gun emplacement upon which ranges, times of flight, times of firing and other information were entered and from which it was possible to deduce data on the future position of the target and the correct time to expose the disappearing gun and fire.

**TOP CARRIAGE** The upper carriage structure of a gun which sits above the trail and supports the trunnions. It can be arranged to recoil along the trail (top carriage recoil, see below), but is more usually arranged to pivot and thus provide traverse to the gun.

**TOP CARRIAGE RECOIL** A system of recoil control in which the gun barrel is attached to the top carriage (see above) without means for recoiling, but the top carriage is allowed to recoil on the bottom carriage, controlled by some form of recoil mechanism. It was very rare and only seen on one or two early railway guns, but the principle was later modified into the 'dual recoil system' (qv).

**TORTOISE** A British self-propelled tank destroyer developed during the latter part of the Second World War. It consisted of a 32pr (94mm) anti-tank gun mounted with limited traverse in the frontal plate of a powerfully-armoured tracked hull. The entire equipment weighed 78 tons, and six were made for trials. The war ended before these were completed and the project was abandoned soon afterwards, as wartime experience had shown the vulnerability of limited-traverse tank destroyers to infantry tank-hunting parties. Four were scrapped, but two remain in the Tank Museum at Bovingdon.

**TRAIL** The part of the gun carriage that extends back from the axle to reach the ground. It acts as a support for the top carriage and resists the tendency for the recoiling gun to pivot round the axle. The trail normally finishes with a spade, which digs into the ground to resist rearward movement, and it may be provided with a firing platform that can be placed on the ground and linked to the trail so that the gun wheels run round the smooth surface of the platform and permit all-round traverse. The side of the trail also forms a convenient anchorage for various pieces of equipment such as tool-boxes, handspikes and other accessories.

**TRAJECTORY** The path described by the shell as it flies through the air. In a vacuum this would be a perfect parabola, but due to the effects of air resistance and gravity it becomes a compound curve, steeper at the target end than at the gun end.

**TRAVELLING LOCK (or CLAMP)** A method of locking the elevating mass of the gun to the trail when travelling. If the gun is not so locked, the motion of bouncing across country will throw enormous stress on the elevating gear, which is the only connection between the carriage and the elevating mass. In some equipments with long barrels it is also desirable to unlock the barrel from the recoil system and retract it to the fully recoiled position, then lock it to the trail, because the long unsupported length of barrel can develop vibrations that throw stress on the elevating gear and, in some cases, can actually induce a permanent bend in the barrel.

**A box trail on a World War One British 6in howitzer.**

**An American 155mm gun in 'march order' showing the barrel retracted and held by one travelling lock under the breech and another under the front of the cradle.**

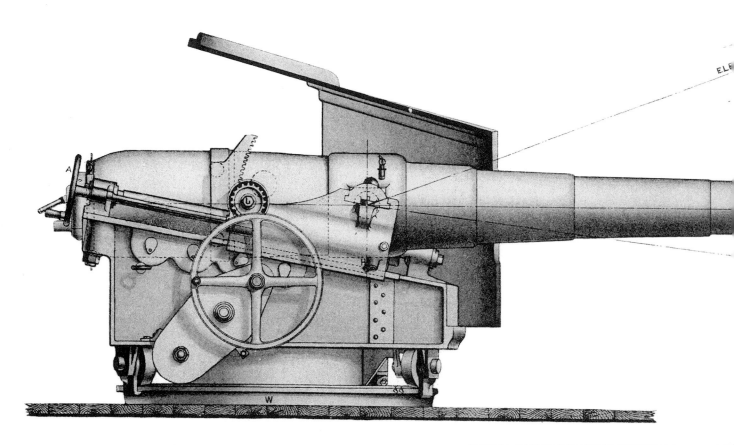

**TRAVERSING GEAR** Apparatus fitted to a gun carriage in order to point the gun barrel in azimuth independently of the carriage itself. It usually consists of a handwheel attached to the trail or lower carriage which operates a shaft on which a worm-wheel is fitted; this meshes with a curved rack on the top carriage so that as the handwheel is turned, the worm-wheel pulls or pushes on the rack and moves the top carriage round its pivot. An alternative method, rarely seen today, is to form the outer surface of the axle into a screw-thread and place a handwheel round it connected to a nut. This hand-wheel forms part of the trail assembly, so that as the wheel is turned it revolves the nut and moves the trail across the axle.

In self-propelled equipments the top carriage may be moved by a rack and worm-wheel, as in towed guns or, in most modern equipments, the gun is mounted in a turret that is revolved by a rack and pinion gear. Power operation can be applied to traversing mechanisms, particularly in self-propelled guns and in artillery of position, but manual operation is always retained as a stand-by.

**TRENCH CANNON** Light artillery for direct fire from trenches; the term is particularly applied to a lightweight 37mm gun developed by the French during the First World War and subsequently adopted by the American Army. It had little application other than in trench warfare, being short ranged and firing a projectile little more destructive than a hand grenade. The 37mm was withdrawn from American service and adapted as a sub-calibre (qv) training device for heavier artillery.

**TRUNNION** Horizontal axles protruding from the side of the gun or cradle and resting in curved bearings in the carriage, allowing the gun to be elevated and depressed. The trunnions pass the recoil force to the carriage, and therefore must be sufficiently strong. They also support the weight of the gun and cradle, and in large equipments this means the adoption of some form of anti-friction mounting so that elevation of the gun is possible. These two requirements are not compatible; the firing stress, hammering the trunnions into their bearings, will damage anti-friction rollers or ball bearings, and therefore it is necessary to devise some form of spring seating that allows the anti-friction system to work when supporting only the weight of the gun, but will retract the anti-friction device when the gun fires and allow the firing stress to be taken on the body of the carriage.

The trunnions are the only part of the elevating component of the gun that stay in the same place, and therefore in some light automatic anti-aircraft guns the trunnions are exceptionally large and designed to allow the ammunition feed to take place through one side, obviating the need to adjust the ammunition feed to the gun's elevation.

Trunnions of old guns are always worth examination; the outer face of the trunnion was often used to record the gun's type, number, maker and date of manufacture.

**TRUNNION RING** A ring shrunk round the gun barrel and fitted with trunnions. It was used to adapt a gun originally manufactured with thrust collars (qv) for use on a carriage requiring trunnions.

**A battery of British 5.25in AA guns in turrets at Gibraltar.**

**TURRET** A revolving armoured structure for mounting a gun so as to permit it to be traversed through a complete circle, or through a lesser arc if restricted by natural features. Although the word is usually associated with the revolving turret on tanks and some self-propelled guns, artillery in turrets has also been widely used in fortresses and in coast defence applications. One of the earliest turret

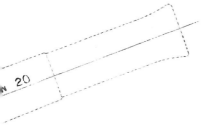

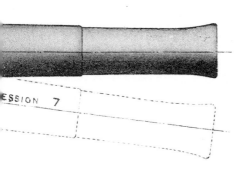

**A British 6in gun on a Vavasseur mounting.**

**VAVASSEUR** A British engineer responsible for a number of inventions related to artillery. The 'Vavasseur Mounting' used a carriage with an inclined top surface; the gun was mounted into a top carriage that could slide on this surface and normally rested at the front, lower, end. The gun unit was connected to the lower carriage through the medium of two hydraulic cylinder assemblies, one on each side of the mounting. One was normally with its piston fully extended, the other with the piston fully compressed, so that when the top carriage recoiled up the slide, one piston was dragged through the oil in its cylinder, the other pushed through; on the return stroke, due to gravity, the piston units acted as brakes to slow the run-out action. This type of mounting was principally used in naval service, though many appeared in light coast defence gun applications.

Vavasseur was also responsible for the perfected design of driving band adopted on British shells from the late 1880s; he received a grant of £10,000 for the use of this invention.

**VENT** The hole running through the breech of a gun through which ignition is carried to the cartridge. In early muzzle-loading guns it was simply a hole drilled through the metal from the top of the gun into the chamber; loose gunpowder was poured into the vent and a flame applied to the exposed end. Later

**Serving the vent during the loading of a muzzle-loading gun, c. 1860.**

designs drilled a much larger hole and inserted a separate metal tube (usually of brass or copper) through which the vent was drilled. This system was adopted because the explosion in the chamber gradually eroded the bottom of the vent and eventually blocked it, and a replaceable vent allowed a complete new 'hole' to be inserted in the gun from time to time.

Ignition was improved in the 19th century by the adoption of various types of 'tube' or 'primer' that were inserted into the top of the vent and fired by percussion or friction to discharge their flame down the vent and into the chamber. It became more practical as guns increased in size to drill the vent into the side of the gun rather than on the top, as a standing man could no longer reach the top.

With the arrival of breech-loading, the vent remained, though it now had to be drilled through the breechblock or screw. In present day bag-charge guns the vent passes through the axis of the mushroom head (qv) and has the slide box and firing lock attached to the rear. The lock is loaded with a primer and fired, and the flash passes down the vent into the chamber just at it always did.

'Serving the vent', in muzzle-loading days, meant stopping-up the outer end of the vent when the gun was being loaded. This practice was adopted in order to prevent air being blown through the vent and possibly inflaming any smouldering fragment of cartridge cloth or powder that might be lodged there. If this happened, the cartridge being loaded would fire, and the man or men on the rammer would be seriously injured or even killed when the rammer was blown out of the gun. The easiest way to serve the vent was to put one's thumb over it, and a split thumb was a hallmark of a gunner in the middle ages. In later years leather thumb-stalls were used, and even mechanical stoppers.

**VENT-PIECE** The breechblock of an Armstrong rifled breech-loading gun, so called because it carried the vent.

installations was for two 16in (40.6cm) muzzle-loading guns on Dover Pier, and although made obsolete in the 1890s the turret, complete with guns, is still in place. Turrets were used by the US coast defence on Fort Drum in Manila Bay, mounting two 14in guns in each of two turrets, and by the Japanese coast defences to mount a number of 40cm guns to cover the Straits of Tsushima in the 1920s.

A self-contained turret and 155mm howitzer developed by Vickers in 1984, a concept that allows obsolescent tank chassis to be quickly turned into modern self-propelled artillery.

An unusual design developed by Vickers in the 1920s was this dual-barrelled gun, one of 44mm for anti-tank fire or an interchangeable one of 60mm for infantry support. The idea was not pursued.

**VERTEX** The highest point reached by the shell on its trajectory. An approximation of the height in feet can be made by squaring the shell's time of flight to the target in seconds and multiplying the result by four.

**VICKERS** The major British armaments company. It began as a small family firm of iron-makers in 1829, and apart from manufacturing iron armour and, later, producing billets of steel for gunmakers, had no involvement with armaments until 1884 when it bought shares in the Maxim Gun Company. This later amalgamated with Nordenfelt (qv) to form the Maxim-Nordenfelt Gun and Ammunition Company, and Vickers held a considerable proportion of the equity. In 1884 Vickers decided to develop steel armour for warships and sell it

abroad. That same year the company was approached by the British government to investigate its potential for manufacturing ordnance. Vickers received its first order for naval guns in 1888, and in the summer of 1890 the first gun passed proof and was accepted for service. Under the guidance of Colonel T.E. Vickers, son of one of the founders, the company set about organizing itself into an entirely self-contained unit for the production of warships, and bought up the Naval Construction and Armaments Company of Barrow-in-Furness in 1897. In the same year Vickers bought out the Maxim-Nordenfelt concern. In later years subsidiaries were acquired or opened in Spain, Italy, Japan and Russia; Vickers set up the Wolseley Tool and Motor Car Company, bought half of William Beadmore of Glasgow, a specialist

armour-plate competitor and, together with Armstrong-Whitworth, assumed control of the Whitehead Torpedo Factory. The company amalgamated with Armstrong-Whitworth in 1927 to form Vickers-Armstrong and in 1930 bought out the Armstrong interests. At present the company is split into two divisions: Vickers Instruments, making optical sights for tanks and artillery; Vickers Shipbuilding and Engineering, the inheritors of the Barrow Works, building ships and the FH70 howitzer; and Vickers Defence Systems, inheritors of the Elswick Works, building armoured vehicles.

Many of the naval guns used by Britain and other countries after 1890 were designed and built by Vickers. Vickers was responsible for, among others, the 13pr and 18pr field guns that armed the British Army from 1904 to 1936, for many of the heavy field and railway guns and howitzers, almost all the British Army's tanks between 1919 and 1938, and in more recent years the 'Abbott' 105mm SP gun and the 155mm FH70 howitzer.

**VICKERS-TERNI** The name given to the Società Italiana d'Artiglieria ed Armamento of Terni. When this company was set up, between 1904 and 1906, Vickers provided about 25 percent of the capital and, in return for a percentage of the profits, supplied the Italian operators with skilled technical knowledge and assistance. The company began building warships for the Italian Navy in 1910 and later went into the manufacture of aircraft. The company also manufactured various Vickers designs of field gun under licence, but this proved to be a relatively minor part of its business. Vickers-Terni went into liquidation in 1921.

# W

**WASHOUT** A water spray system built into the breech of a heavy gun so that water can be injected into the chamber before the breech is opened. This extinguishes any smouldering fragment remaining from the bag-charge cartridge. If such a fragment were not extinguished, the draught caused by opening the breechblock might fan it into flame and thus present the danger of igniting the fresh cartridge when loaded or of flashing back from the chamber and igniting the cartridge waiting to be loaded.

**WATERVLIET ARSENAL** The principal American government gun-making facility, situated in New York state, on the Hudson River just north of Albany. Founded in 1812 it became the principal factory for the manufacture of cannon in 1889, a position it has held to the present day.

**WEAR** In the artillery context, wear means the gradual process of wearing away of the interior surface of the bore as a consequence of shooting. There are two types of wear in guns, abrasive and erosive. Abrasive wear is the wear caused by the actual friction generated by the passage of the shell up the bore, and is relatively small. Erosive wear, which is by far the more destructive, is caused by the hot gases generated by the explosion of the cartridge. The gas blows past the shell driving band in the first micro-seconds of firing, before the driving band is fully seated into the rifling, and because the explosion temperature is somewhat higher than the melting point of most steels, this begins to 'wash away' the surface of the bore at the commencement of rifling. Once erosion begins, it accelerates, as the initial path is soon enlarged by more and more gas flowing through it. Condemnation of guns is often governed by the amount of erosive wear at the front of the chamber, because experience has shown that such erosion is directly linked to loss of velocity, range and accuracy. The amount of erosion is often proportional to the number of full charges fired, so the gunner keeps a record of 'Effective Full Charges', assumes a certain degree of wear, and adjusts his sights or ballistic computations for the assumed loss of velocity proportional to the assumed wear. See 'EFC'.

**A severe case of erosive wear in a 57mm gun. The rifling has almost vanished, leading to instability of the shot, which has in turn left a bulge in the bore.**

A Whitworth 6pr gun, with its peculiar twisted-hexagon projectile beneath it

**Left: A Welin breech screw with the mushroom head and obturating pad removed.**

**WELIN SCREW** A type of breech screw in which the circumference is stepped in three different diameters; when rotated so that it is free to open, the reduced diameters permit the screw to begin swinging open immediately without having to be withdrawn axially from the breech aperture.

**WESTERVELT BOARD** The name given to a Board of Officers convened by the US War Department on 11 December 1918 to 'make a study of the armament, calibers and types of matériel, kinds and proportions of ammunition, and methods of transport of the artillery to be assigned to a field army'. The name was derived from Brigadier-General William I. Westervelt, senior officer of the Board; it has sometimes been referred to as the 'Caliber Board'. The necessity for this enquiry arose because of the unpreparedness of the US Army for the First World War and its lack of knowledge of up-to-date European artillery technology and tactics when it entered the war; the purpose of the Board was to examine all European artillery and make recommendations for the future. The Board reported on 5 May 1919 and subsequent field artillery development in the US Army largely followed its recommendations. The report was printed in the *Journal of the US Artillery*, Vol 51, No 1, July 1919. Contrary to common belief, the Westervelt Board was not concerned with armoured vehicles and made no recommendations about tanks.

**WHITWORTH** Sir Joseph Whitworth (1803-87) was a prominent British mechanical engineer whose principal fame rests upon his development of precise measuring systems and the standardization of screw-threads and gauges. In the 1850s he put forward a design of infantry rifle that was turned down by the War Office, and he then began developing cannon, principally in an endeavour to oust Armstrong from his place of prominence. The two unique features of Whitworth's design were firstly the method of spinning the projectile, done by forming the interior of the barrel into a twisted hexagon. The projectile was a cylinder planed on six sides to form an hexagonal section; this was a mechanical fit in the bore, so that as it passed up the bore the twist of the hexagon gave rotation to the shot. The second feature was his method of closing the breech, which used a cap with interrupted screw-threads that went over the exterior of the barrel, rather than locking into threads in the barrel as did other systems. Whitworth made muzzle- and breech-loading guns, which were frequently fired in trials against Armstrong and other designs, but they were never adopted by the British services, though a number were sold abroad, particularly to the Confederate forces during the American Civil War. The general opinion seems to have been that the Whitworth gun was accurate when it worked well, but there was always the danger of the hexagonal shot wedging itself in the bore and damaging the gun. After Whitworth's death the company was incorporated into Sir William Armstrong's firm to become Armstrong-Whitworth.

**WINDAGE** The difference between the external diameter of a projectile and the internal diameter of the weapon barrel. There must obviously be some difference otherwise the projectile would be an 'interference fit' and would not be able to pass along the bore, but the amount can be critical to accuracy and efficiency in most types of weapon. In modern breech-loaded equipments it is less vital because the escape of gas past the projectile is prevented by the driving band, and provided the windage is not so excessive to allow the shell to rattle from side to side the weapon will operate well enough. In muzzle-loaded weapons, however, whether we are referring to 17th century cannon or modern trench mortars, the windage becomes extremely important; it is necessary to have a fairly generous windage in order to allow the projectile to be loaded, otherwise air would be trapped beneath it and would prevent the projectile reaching its proper

**Daily maintenance on Winnie, the 14in gun that fired across the English Channel; Dover, 1942.**

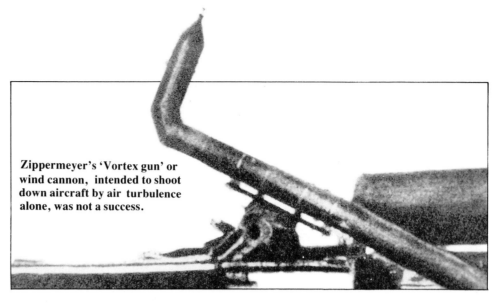

Zippermeyer's 'Vortex gun' or wind cannon, intended to shoot down aircraft by air turbulence alone, was not a success.

place in the bore. On the other hand, too much windage means a loss of propelling gas round the projectile when it is fired, irregular movements of the projectile from side to side as it goes up the bore, and consequent loss of accuracy and range.

**WIND CANNON** Developed by a German theorist called Zippermeyer during the Second World War, the Wind Cannon was a steel cylinder with a nozzle inside which a mixture of oxygen and hydrogen ignited. The subsequent explosion blew a jet of air from the nozzle with extreme force, and it was claimed that on trials the jet blew a hole in a 1in pine plank at 200 metres (219 yards) range. It was hoped that this could be used an an anti-aircraft weapon, the jet either damaging the aircraft or, at least, setting up vortices and eddies in the surrounding air that would make the aircraft unstable and cause it to crash. One gun was built and installed to protect a bridge near Stuttgart, but it proved quite useless and was later scrapped.

**WINNIE AND POOH** Two 14in (35.5cm) guns installed near Dover, in south-eastern England, in 1940 for the purpose of bombarding German batteries in France. The idea was proposed by Winston Churchill, and two guns belonging to the reserve stocks for the battleship *King George V* were fitted to modified naval mountings and installed in concrete emplacements, manned by Royal Marines. The first gun was in place by August 1940 and was christened 'Winnie' upon its inspection by Churchill. The second gun was installed in February 1941 and was christened 'Pooh', an allusion to the children's book *Winnie the Pooh*. During the subsequent year both guns frequently fired in the counter-bombardment role against the German batteries on the French coast. The guns were 16.5 metres (54.1ft) long, could elevate to 55° and had 65° arc of fire each side of a central line. The shell weighed 720kg (1,578lb), had a velocity of 750m/sec (2,461ft/sec) and a maximum range of 43.2 kilometres (27 miles).

**WIRBELWIND** German self-propelled anti-aircraft gun equipment consisting of a four-barrelled 20mm cannon unit placed in a rotating six-sided turret on a Panzer IV chassis. Intended to supplement production of the 'Möbelwagen' (qv) as a mobile air defence equipment for the protection of armoured columns, 86 were built in late 1944, but production ceased when it was found that the 20mm guns were less effective than the twin 3.7cm guns used on Möbelwagen.

**WIRE WOUND GUN** A gun manufactured by wrapping several miles of steel wire round the barrel in layers under tension. This compresses the 'A' Tube and acts in a similar manner to shrinking a hoop on top of the tube to place the 'A' Tube under compression. The wire is retained in place by a hoop or jacket placed above it. The advantage is that the wire can be tested thoroughly, whereas a hoop could have a flaw in it. The system was developed in the 1890s when hoop construction was a doubtful manufacturing proposition, and went out of use when auto-frettaging came into service. Another advantage was that a wire-wound gun was usually lighter than a built-up gun of similar strength, but a disadvantage was that the wire winding gave no longitudinal strength to the gun, which often led to droop.

**WITNESS POINT** A gunnery technique for determining the range and azimuth to a target without firing ranging shots at it. A 'witness point' of approximately the same range was selected from the map and the gun ranged on to it. Comparison of the range and azimuth deduced from the map and the range and azimuth actually fired to obtain a hit gave a difference that represented corrections due to meteorological conditions and changes in the gun's performance from standard. These corrections were then applied to the map data for the target and the gun fired on this corrected data to obtain a hit with the first shot. Witness point procedure was used in the days when 'meteor' data was either not available or of dubious accuracy, and is no longer used.

**WOOLWICH ARSENAL** Properly known as the Royal Arsenal, Woolwich was the principal British government gun manufactory from the 17th century until the Second World War, when gunmaking was transferred to other Royal Ordnance Factories. Situated on the River Thames below London, Woolwich was divided into three principal units concerned with artillery: the Royal Gun Factory, the Royal Carriage Department and the Royal Laboratory, the latter being responsible for the development and manufacture of ammunition. Woolwich virtually had a monopoly on gun design and manufacture, though some manufacture was contracted out, until the advent of the Armstrong gun (qv), which was designed and initially produced by a private company. With the rise of commercial gunmakers such as Armstrong and Vickers, more of the gunmaking was contracted out, though Woolwich still acted as the design authority and development agency and performed a proportion of manufacture. Woolwich Arsenal was closed in the late 1960s.

**WOOLWICH INFANT** The name given to the 12in rifled muzzle-loading gun of 35 tons developed at Woolwich Arsenal in 1871. It was originally built in 11.6in calbire, but was then enlarged to 12in to give better combustion of the propelling charge. Of the 15 built, most were used as naval turret guns, though guns were emplaced as coast defence guns. The guns fired a 316kg (697lb) piercing shell at 415m/sec (1,362ft/sec), using a 50kg (110lb) charge of black powder.

# YZ

**YAW** The deviation of the axis of the shell from its theoretical trajectory during flight. It is a complex movement and can best be simplified by considering it in two phases; immediately after leaving the muzzle, and during the curved part of the trajectory. In the former case some yaw develops because of the loose fit of the shell in the bore and the slight disturbances due to the rush of gas around it as it leaves the muzzle. This initial yaw is relatively slight in a rifled weapon, but can be considerable in the case of smoothbored weapons firing fin-stabilized projectiles. Later in the flight the shell, gyroscopically stabilized, tends to maintain its flight attitude while the trajectory it is following adopts a curve, with the result that the nose of the shell deviates from the axis of the trajectory. This allows air to strike the nose and gives a deflecting effect that has a tendency to push the shell even further from the trajectory. Any irregularity in the balance of the shell due to faulty manufacture will set up a centrifugal force that will add to the yaw.

**YOKE MOUNTING** A special type of gun mounting developed for the 12in breech-loading guns installed in the iron forts in the Solent in the 1880s. These guns were to be installed in casemates that had been designed for muzzle-loaders, and there was insufficient room to permit the normal type of carriage and recoil system to be used. The Yoke Mounting was an iron casting anchored to the floor and roof of the casemate, close to the gun port, and had a hole in the middle through which the gun barrel passed and to which hydraulic bufers were attached.

**ZALINSKI** A lieutenant of the US Coast Artillery, Edward L. Zalinski was involved in the initial trials of Mefford's 'Dynamite Gun' (qv) in 1884. After the trial, Mefford went back to make some im-

provements, but while he was so engaged a mysterious Mr. G.H. Reynolds stepped in and took out patents covering Mefford's unpatented ideas. Zalinski retired from the Army and joined Reynolds in forming the Pneumatic Dynamite Gun Company with the result that the weapon became generally known as the Zalinski Dynamite Gun and Mefford was forgotten.

**ZONE OF THE GUN** The rectangle round the target into which the shells fired from a gun will fall. For a given elevation, a series of shells will not fall in the same spot but will be spread around the theoretical impact point in accordance with the laws

**Woolwich Arsenal in its prime; the howitzer barrel shop in 1917.**

of distribution and probability. The 50 percent zone is that rectangle into which 50 percent of the shells will fall, and is tabulated in the firing tables for every 100 yards (91 metres) of range. The 100 percent zone is four times the 50 percent zone and is that area in which all the shells will fall. What all this means in practice is that with the best information and equipment, it is still impossible to guarantee hitting a precise spot, and the gunner must have the zone of the gun in mind when ranging fire on a target.

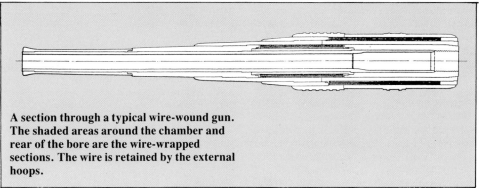

**A section through a typical wire-wound gun. The shaded areas around the chamber and rear of the bore are the wire-wrapped sections. The wire is retained by the external hoops.**

# INDEX

# ACKNOWLEDGMENTS

The illustrations in this book are from the author's personal collection, except for photographs reproduced by permission of the following sources:

Bodleian Library, Oxford p.10 top, p.13 bottom; Groupment industrielle d'armaments terrestre, Paris p.56 inset, p.60 lower; A.T. Hogg p.57 bottom, p.58 bottom left, p.60 upper, p.63 upper, p.64 upper; Imperial War Museum, London p.51; Mansell Collection, London p.17; Mary Evans Picture Library, London p.7 top; Military Archives and Research Services, London p.8 top left, p.35 top, p.38, p.42 bottom; National Army Museum, London p.37 top; Personality Picture Library, London p.15; Southern Bye-Train of Artillery, London p.14 bottom; G.Z. Trebinski p.58 top and bottom right, p.60 lower.

Thanks are due to the Commandant and the staff of the Guns Department, Royal Military College of Science, Berkshire, UK, for their permission to photograph various details of equipments; to Anna Teresa Hogg for the photographs taken at Aberdeen Proving Ground, Maryland, US; to the Royal School of Artillery, Wiltshire, and the Royal Artillery Museum, London, for photographs taken at those establishments at various times over the past forty years.